SpringerBriefs in Fire

Series Editor
James A. Milke

For further volumes:
http://www.springer.com/series/10476

Erdem A. Ural

Towards Estimating Entrainment Fraction for Dust Layers

 Springer

Erdem A. Ural
Loss Prevention Science
and Technologies, Inc.
Stoughton, MA, USA
erdem.ural@lpsti.com

ISSN 2193-6595   e-ISSN 2193-6609
ISBN 978-1-4614-3371-2   e-ISBN 978-1-4614-3372-9
DOI 10.1007/978-1-4614-3372-9
Springer New York Dordrecht Heidelberg London

Library of Congress Control Number: 2012931562

Printed on acid-free paper

Springer is part of Springer Science+Business Media (www.springer.com)

Preface

This report describes the work performed in the first phase of the Fire Protection Research Foundation project entitled Dust Explosion Hazard Assessment Methodology.

Dust explosions occur only when a number of preceding events take place almost simultaneously. A comprehensive generic chain of events that can lead to an explosion is described in Chapter 17-8 of NFPA Handbook of Fire Protection.

Unless dust is kept suspended in the air by design, most dust explosions start with a disturbance that raises dust into suspension. The disturbance could be as simple as the rupture of a compressed air line, a mechanical jolt to beams where dust layers have accumulated, or a small explosion somewhere else in the plant. Once subjected to the disturbance, the amount of dust entrained (removed) from the deposit depends predominantly on the magnitude and the severity of the disturbance. Other parameters such dust and layer properties can also play a role on the dust entrainment rates.

Once the dust is lifted off from the layer, it mixes with air and can form explosible pockets in the enclosure. The size of the explosible cloud volume is controlled by the dust entrainment rate and the existing air motion in the enclosure (e.g. ventilation/recirculation), or that induced by the primary disturbance.

Once created, an explosible dust cloud can be ignited by a number of possible ignition sources. In critical applications or for hard to ignite dusts, credible strength of ignition sources can be evaluated and compared to the ignition requirement of the particular dust cloud. Though seldom done, such an exercise may reveal whether elimination of ignition sources is a viable prevention method for the particular application. While NFPA standards encourage ignition control to reduce the frequency of the incidents, they generally assume that an ignition source may exist despite the presence of an ignition source control program.

Once ignition takes place, the reaction front (flame) moves into the unburnt dust cloud with a well-defined velocity, called the burning velocity. If the enclosure is practically unvented, the maximum explosion pressure is related to the heat of combustion of the dust cloud. Fully confined deflagrations of dust clouds occupying a substantial portion of the enclosure volume commonly develop pressures in the range of seven to ten times the initial absolute pressure, or 100 to 140 psig (7 to 10 barg). If the enclosure has large openings (deflagration vents), then the maximum explosion pressure is substantially reduced depending on the burning velocity and the maximum flame surface area. Pre-existing turbulence conditions inside the enclosure, turbulence induced by flame propagation, and the geometry of the enclosure can increase both the burning velocity and the maximum flame surface area.

Process enclosures are seldom designed for pressure containment. Pressures at which enclosure failure occurs can be quite low, particularly for enclosures of rectangular sheet metal construction. These can fail completely at internal pressures of a few pounds per square inch. Typically, buildings can tolerate only a fraction of 1 psi pressure.

While properly designed deflagration vents can successfully limit the peak pressure rise to a level that can be tolerated by the enclosure, an explosible dust cloud occupying a substantial portion of the compartment volume is never allowed in occupied enclosures, because it is capable of producing untenable conditions in the entire volume. Combustible dust occupancy standards promulgated by NFPA recognize this fact and impose restrictions on dust accumulations, currently specified as the threshold layer thickness and areas. For example, the current (2006) edition of NFPA 654 implies a threshold layer depth of 1/32 inch while NFPA 664 uses a layer depth of 1/8 inch. NFPA 654 permits adjusting the layer depth criterion for variations in dust bulk density while NFPA 664 does not.

The objective of this project is to establish the technical basis for quantitative criteria for determining that a compartment is a dust explosion hazard that can be incorporated into NFPA standards or other relevant safety codes. For the purpose of this study, a dust explosion hazardous condition is defined as that which creates a hazard for individuals and property, which are not intimate with the initiating event. The scope of the first phase of the project is limited to a study of those combustible dusts covered under the scopes of NFPA Standards 61, 484, 654 and 664 which include dusts encountered in agricultural and food processing, combustible metals, wood processing and wood-working facilities. However, since these standards cover dusts exhibiting a wide spectrum of properties, the project results could be extrapolated to most other dusts. In fact, large scale test data for coal dust rock dust mixtures as well as sand and soil were used in the preliminary validation of the strawman method described in this report.

As apparent from the objective, the biggest challenge in a project of this sort is to simplify the models to an extent that would be suitable for incorporation in NFPA Standards and Codes. This is a difficult task to accomplish for two reasons:

1) the level of complication that is suitable for incorporation in NFPA Standards is at best a subjective concept, and
2) simplification often comes at a cost of loss of generality, and added conservatism for some applications.

Many valuable discussions with the project panel helped the author develop a strawman method which provides a good balance between the two desirable but competing features: simplicity versus generality.

Acceptable simplicity was achieved by assuming all entrained dust enters into a dust cloud, which is always at the worst-case concentration for the particular combustible dust. The more dust is entrained, the bigger the cloud is. This assumption is conservative but obviates the need for complex tools such as computational fluid dynamics to calculate the development of the entrained dust cloud. This assumption also allows ready use of the partial volume deflagration concepts and equations provided in NFPA 68, a published consensus standard.

Hence, the strawman method is essentially reduced to two components:

a. selection of the types, magnitudes and durations of the maximum credible disturbances
b. calculation of the mass of the dust entrained from the deposits.

The first component depends on the primary event scenarios that are credible for the specific occupancy and are expected to evolve through the consensus process. To demonstrate the concepts in this report, two most common scenarios were selected after conferring with the project panel.

First scenario is the catastrophic burst of indoor equipment. Resulting blast wave propagates over the dust layer and raises all of it or a portion of it into suspension. A worked out example included in the report demonstrates how the amount of dust lifted from the layer can be estimated by relying on published pressure vessel burst nomograms to calculate the magnitude of the air velocity pulse and its duration.

The second scenario is the deflagration venting from a room into a building covered with combustible dust deposits. In the worked out example included in the report, the flow field induced by the vent discharge is approximated by using axial jet correlations, and its duration is calculated using an equation provided in NFPA 68.

The second component of the strawman method is the calculation of the mass of the dust entrained from the deposits. An extensive international literature review was carried out on relevant research on airflow induced dust entrainment rates. Effects of factors such as aerodynamic flow and boundary layer characteristics, dust particle size and shape, and dispersibility were examined. Since the dust entrainment occurs deep in the boundary layer, friction velocity rather than the free stream velocity is the more appropriate parameter to correlate the entrainment rate. On the other hand, most users of the NFPA standards are not anticipated to be versatile in using aerodynamics concepts encompassing the friction velocity. Therefore, an additional simplification is introduced by translating the selected entrainment rate correlation to free stream velocity. The selected equation was also modified for low flow velocities so that entrainment rate tends to zero at the threshold velocity.

The following equation is proposed to estimate the entrainment mass flux[1] until the validation tests are completed in the next phase of this project:

$$m = 0.002 * \rho * U * \left(U^{1/2} - U_t^2/U^{3/2} \right) \quad U > U_t \tag{1}$$

---

[1] The rate of mass removal per unit area per unit time.

where:

m'' entrained mass flux in kg/m2-s
ρ gas density in kg/m3
free stream velocity in m/s
U_t threshold velocity in m/s.

The threshold velocity, U_t, is the minimum air velocity at which dust removal from the layer begins, and it depends on factors such as particle size, particle shape and particle density. The report provides algebraic correlations and charts to estimate this parameter.

Predictions of the strawman method is compared to available large scale coal dust and cornstarch explosion test data. Good agreement was observed. Nevertheless, additional tests are recommended to validate equation (1) further.

The new strawman method described in this report represents a paradigm shift in dust explosion hazard assessment. The approach used in current NFPA standards implicitly assumes that the dust explosion hazard is primarily related to the thickness of the dust layers, or the total mass of the dust accumulations. The new strawman method, on the other hand, primarily determines the maximum amount of dust an initial disturbance can raise into a cloud. If that quantity is large enough to create an explosion risk, then the explosion hazard can still be averted by controlling the amount of dust accumulations.

In other words, the new strawman method is capable of estimating the fraction of the dust accumulations that can become airborne, a parameter also known as the entrainment fraction. Predicted entrainment fraction values range anywhere from zero to one, depending on a number of parameters including dust characteristics, layer thickness, geometry, as well as type, magnitude and the duration of the maximum credible disturbance.

Acknowledgements

This work was made possible by the Fire Protection Research Foundation (an affiliate of the National Fire Protection Association). The author is indebted to the companies who provided the funding for this project through the Research Foundation, and to the project technical panel members for many valuable suggestions. Special thanks go to Kathleen Almand, Henry Febo, Walt Frank, and Sam Rodgers for serving on the special task group. The author is also grateful to Marcia Harris of NIOSH and Michael Sapko of Sapko Consulting Inc. for many valuable discussions, and for making unpublished NIOSH data available to him.

Contents

Editor's Note

The equations analyses and the conclusions presented in this report are based on a review of the documents available at the time this report was prepared. The findings and conclusions in this report are those of the author and do not necessarily represent the views of the Fire Protection Research Foundation or the National Institute for Occupational Safety and Health.

List of Contributors

Project Technical Panel

Panel Members
Elizabeth Buc, Fire and Materials Research Laboratory
Henry Febo, FM Global
Walt Frank, Frank Risk Consulting
Paul Hart, XL Insurance
Joe Senecal, Kidde Fenwal
Guy Colonna, NFPA staff liaison

Project Sponsor Representatives
Brice Chastain, Georgia Pacific
Greg Creswell, TIMET
Mark Holcomb, Kimberly Clark
Robert Hubbard, Abbott Laboratories
Jerry Jennett, Georgia Gulf Sulfur
David Oberholtzer, Aluminum Association
George Olson, Procter and Gamble
Sam Rodgers, Honeywell

List of Figures

Chapter 1
Introduction

The objective of this project was originally specified as to "perform literature review on relevant research and dust explosion incidents focused on those factors which impact the dust hazard assessment, such as dispersibility (entrainability), layer thickness and entrainment characteristics of dust particles, facility geometry and deposition characteristics, etc."

During the teleconference held on October 6, 2009, project panel has reviewed the task objective and refocused it on research and testing, not fire/explosion incidents. NFPA 654 ROC (A2010 published methodologies to determine maximum allowable dust accumulations or minimum cleaning frequencies. Published formulas rely on an a priori value for the dust entrainment fraction selected to provide a level of protection, for typical occupancies, comparable to that implied by the previous editions of NFPA 654. This Research Foundation project focuses on collecting available information which may be useful to NFPA committees in making informed decisions about the appropriate value of dust entrainment fraction.

Other tasks of the authorized phase of the project focuses on the development of a proposed strawman dust explosion hazard assessment method based on those parameters which, if validated, would be suitable for incorporation in NFPA Standards and Codes.

One of the necessary conditions for the occurrence of dust explosions is the dispersion of combustible dust in air. In industrial situations, dust dispersion could be (Hertzberg, 1987):

(1) an integral part of the process, as in a pulverizer, or a pneumatic transport line;
(2) a by-product of the process, as in dust handling equipment; or
(3) caused by an accident such as a ruptured compressed air line, or a blast wave emanated from a nearby explosion.

In the first category, efficient dispersion of dust is wanted, in the second category, the dust dispersion is considered a nuisance, mostly from the industrial hygiene point of view, and suppression techniques such as wetting or oil mist

E.A. Ural, *Towards Estimating Entrainment Fraction for Dust Layers*, SpringerBriefs in Fire, DOI 10.1007/978-1-4614-3372-9_1,
© Fire Protection Research Foundation 2011

treatments[1] are employed. The types of accidents included in the third category may suspend large quantities of combustible dust in air for relatively short periods, and may lead to severe dust explosions. For suitable plant geometry and fuel distribution, even a mild primary dust explosion may lead to cascading "secondary explosions" in which the aerodynamic disturbance caused by the primary explosion lifts the dust originally deposited on surfaces and mixes it with the air, thereby creating additional paths of flammable dust-air mixture for the flame to travel. Multiple secondary explosions are not uncommon in industrial dust explosions and can be responsible for severe losses. For example, based on detailed investigations of fourteen grain elevator explosions occurring between January 1979 and April 1981, Kauffman (1987) attributes, on the average, 85% of the fatalities, 89% of the injuries, and 96% of the property loss to secondary explosions, with primary explosions making up the small balance. Similar conclusions are drawn from the recent CSB investigations.

Accumulations of dust inside enclosures are normally found on the floor, as well as on other surfaces such as beams, equipment and structural elements. The elevation of the surfaces covered by dust layers can be expected to affect the dust cloud size and concentration, since the gravitational force tends to reduce the dispersion, if the dust accumulations are close to the floor, and aid it, if they are close to the ceiling. In the first case, the energy to overcome gravity must be supplied by the disturbance.

The disturbance created by the primary explosion could be of aerodynamic nature or in the form of vibrations transmitted by solid structures. While vibrations can be responsible for dispersal of some of the dust located near the top of an enclosure or components possessing just the right degree of stiffness and mass, most dust is usually dispersed by the aerodynamic disturbance. This disturbance is of the transient type and can last anywhere from a fraction of a second to several seconds. For a secondary explosion to be possible, the disturbance must be of sufficient strength and duration to:

1. remove dust particles from the layer;
2. mix the dust with air to form a flammable (explosible) dust cloud; and
3. prevent the dust cloud from settling until it is ignited by local ignition sources or by the arrival of a flame front propagating from the primary explosion.

Hence, the primary focus of this project is on the removal by aerodynamic disturbances of dust particles or agglomerates from layers or piles of cohesive and non-cohesive dusts of varying particle shapes and densities.

---

[1] These additives cause an increase in the cohesion of the particles in the layer, thereby requiring stronger disturbance for their entrainment. Treated dust also tend to peel as large agglomerates rather than individual particles, which tend to settle out faster. Dust abatement techniques of this type have been in use in a number industries such as grain, coal and chemical.

Chapter 2
Literature Review

2.1 An Overview of Aerodynamic Entrainment of Dust Particles into Air

Aerodynamic forces acting on a dust layer can dislodge particles or clumps of particles from the layer and set them into motion. The entrainment of dust layers occurs in various modes or their combinations. Powders demonstrating negligible cohesion[1] tend to be removed as individual particles. More cohesive powders are removed as groups of particles (agglomerates), and sometimes, depending on layer and surface properties, appreciable portions of the layer can be lifted as a whole.

When the entire layer is subjected to uniform aerodynamic conditions (as in the case of pipe flow, or atmospheric flow) the dust may either be removed uniformly over the entire layer (erosion), or may be removed from the leading edge of the deposit. In this latter process called denudation, the leading edge of the deposit propagates in the direction of the flow. It is generally believed that for erosion type dust removal, adhesive forces must be larger than the cohesive forces.

As seen in Fig. 2.1, the particles removed from the layer can also show different types of behavior. Dislodged particles may roll on the surface until they find a spot with reduced fluid forces (such as a pit) and come to rest, or collide with another particle, thus transferring their momentum and aiding the removal of the new particle. During their travel, particles may even become airborne for short periods of time, yet still remain close to the surface. This type of transport is called surface creep. In another mode of transport called saltation, the particles are ejected from the surface almost vertically and are carried by the wind horizontally until they fall back onto the surface. Bagnold (1941) observed saltation layer thicknesses in the order of meters for desert sand. At higher air velocities, particles

---

[1] Conventionally, the word cohesion refers to the attraction force between two surfaces of the same material (such as the dust particles), whereas adhesion implies different materials (such as dust particle attracted to a plate)

E.A. Ural, *Towards Estimating Entrainment Fraction for Dust Layers*,
SpringerBriefs in Fire, DOI 10.1007/978-1-4614-3372-9_2,
© Fire Protection Research Foundation 2011

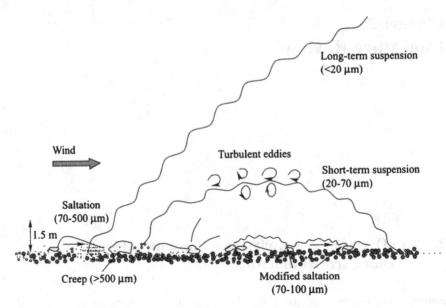

Fig. 2.1 Different types of sand particle motion that can occur during wind erosion (from Shao, 2008)

do not return to the surface (referred to as entering into suspension) and are carried for long distances.

Intuitively, it is easy to see whether a particle will be removed from a surface or from another particle, and the number of particles that get removed per unit time per unit surface area are governed by a balance between the forces trying to dislodge the particle, and the forces trying to keep the particle in place.

2.2   Characterization of the Aerodynamic Forces Acting on a Solid Particle

Extensive reviews of this subject are given by Clift (1978), and Yoshida et al (1979). Conventionally, aerodynamic force is resolved into two components: drag, in the direction of the mean flow; and lift, perpendicular to it. For uniform flow over a particle, the drag force is expressed by

$$F_D = CDA1/2\rho U^2 \qquad (2.1)$$

where C_D, A, ρ, and U, respectively represent the particle drag coefficient, particle cross—sectional area, air density, and air velocity with respect to the particle.

At low relative velocities, the drag coefficient decreases with increasing Reynolds number. For spherical particles, an approximation to drag coefficient for Reynolds numbers smaller than 1000 is given by (Clift, 1978):

$$C_D = 24/Re\left(1 + 0.15Re^{2/3}\right) \qquad (2.2)$$

where:

Re $\rho UD/\mu$ particle Reynolds number;
D particle diameter;
μ viscosity of air.

The first term in this equation is called the Stokes drag coefficient, which constitutes the major portion of the drag for small particles. The corresponding Stokes drag can then be calculated as:

$$F_{Dst} = 3\pi\mu UD \qquad (2.3)$$

A classical application of uniform flow over a sphere is the free fall of spheres. The terminal fall velocity is calculated by equating the drag force to the particle weight. In the Stokes regime (Re $<<$ 1) this is given by:

$$U_t = \rho_p g \ D2/(18\mu) \qquad (2.4)$$

where ρ_p is the true density of the particle.

At high particle Reynolds numbers, the drag coefficient becomes independent of the Reynolds number so that the drag force is proportional to the square of both the relative velocity and the particle size, whereas the terminal velocity is proportional to the square root of the particle size and density.

The terminal velocity of an ensemble of monodisperse spherical particles in the absence of agglomeration is lower than that measured for an individual particle. This hindered settling problem has been studied in detail for monodisperse and polydisperse suspensions. Examples of such work can be found in Batchelor (1972), Batchelor and Wen (1982), and Davis and Birdsell (1988). The correction in the terminal velocity due to hindered settling is of the same order of magnitude as the volumetric fraction of the solid particles and should be negligible for most explosible dust clouds.

The drag force for low Reynolds number shear flow around a spherical particle is usually calculated by assuming the relative flow velocity as the undisturbed value at the particle center. This is rigorously accurate only for small particles exposed to constant velocity gradient because the Stokes drag force is proportional to the relative velocity. If the particle is adjacent to a wall, however, the drag force can be 70% higher than the Stokes drag, as suggested by the creeping flow solution of O'Neill (1968):

$$F_{Dw} = 8\mu\gamma D^2 \qquad (2.5)$$

where γ denotes the velocity gradient perpendicular to the wall. The effect of flow shear on particle drag is more difficult to assess at high Reynolds numbers due to flow separation phenomenon.

The lift force on a spherical particle is induced if the particle is rotating (axis of rotation perpendicular to the direction of flow) or is subjected to shear flow. Most work on this topic is concentrated on either very low or very high Reynolds numbers.

The lift force on a sphere spinning in a uniform flow was calculated for small Reynolds numbers by Rubinow and Keller (1961) as:

$$L = 1/8\pi\rho D^3 U\omega \tag{2.6}$$

where ω is the angular spin velocity. It is interesting to note that this solution is independent of viscosity, and is in a form similar to the Kutta—Joukowsky formula used to predict lift due to potential flow.

Saffman (1965) calculated a lift force exerted on a spherical particle by a shear flow. The formula he developed for low Reynolds numbers:

$$L = 1.61\rho\nu^{1/2}\gamma^{1/2}D^2 U \tag{2.7}$$

where γ is the magnitude of the velocity gradient. Saffman's analysis has shown that up to the maximum spin attainable by free particles due to shear, the effect of angular velocity on the lift force is of higher order than the that calculated from the above equation. The velocity, U, in this equation is taken as the undisturbed velocity at the particle center. For a particle resting on a flat surface $U = \gamma D/2$, therefore:

$$Lw = 0.8\rho\nu^{1/2}\gamma^{3/2}D^3 \tag{2.8}$$

Another small Reynolds number particle lift mechanism was postulated by Cleaver and Yates (1973) for turbulent boundary layers. This model is based on the turbulent burst phenomenon occurring as sudden random eruptions in the boundary layer, transporting fluid near the wall towards the mean flow. Treating bursts as viscous stagnation flow, and somehow estimating the strength of the stagnation flow from the measurements of the mean velocity fluctuations normal to the wall, Cleaver and Yates (1973) proposed the following formula for particle lift due to turbulent burst:

$$L = 0.076\rho\nu^{1/2}\gamma^{3/2}D^3 \tag{2.9}$$

which is an order of magnitude smaller than the Saffman's lift.

The lift due to particle spin in uniform flow at high Reynolds numbers, called the Magnus force after its inventor, is well known to many tennis and golf players. The magnitude of this force is determined experimentally (see e.g., Clift (1978)), since the lift is generated by the formation of an asymmetric wake.

A sphere resting on a flat plate also experiences a lift force, as well as increased drag due to the presence of the wall, at high Reynolds numbers. Okamoto (1979) measured a lift coefficient $C_L = 0.242$, and drag coefficient $C_D = 0.627$ at Re = 4.74×10^4.

The free stream turbulence is expected to have an important effect in the dispersion of a dust cloud once the particles are removed from the surface because it may control the extent of dispersion, settling rate and agglomeration/ deagglomeration phenomena. A review of free stream turbulence effects on single particle behavior is given by Clift (1978). The particle dispersion by turbulence field is believed to be strongly dependent on the Stokes number, which is defined as the ratio of the characteristic particle response time to the time scale of the large scale eddies. The characteristic particle response time can be estimated as the ratio of terminal velocity to gravitational acceleration. At low Stokes numbers (i.e., particles with small settling velocity), the particles faithfully follow the fluid motion, and they are dispersed at approximately the fluid diffusion rate. At large Stokes numbers there is hardly any particle dispersion. Interestingly, there is an intermediate Stokes number regime in which particles may be dispersed faster than the fluid would, believed to be due to particles actually flinging out of eddies (e.g., Chein and Chung (1988)).

At intermediate—to—high Stokes numbers, mean particle drag may be decreased or increased due to presence of turbulence. Transition of flow around particle to turbulence is known to sharply reduce the drag coefficient. The presence of free stream turbulence causes this transition to occur at lower Reynolds numbers than critical. One of the mechanisms of increased particle drag is observed beyond the Stokes regime where the particle drag is a stronger than linear function of relative velocity. As a result of this functional dependence, a particle subjected to sinusoidal velocity fluctuations superposed on a mean flow should experience a larger increase in drag force during positive phase compared to the decrease in drag during the negative phase. When averaged over the cycle, therefore, a net increase in drag arises. As a result, a decrease in terminal velocity is observed in a fluctuating flow field. If the velocity fluctuations are of sufficient strength, the average terminal velocity reaches zero. The former phenomenon, called levitation, has been observed for solid particles suspended in liquid (Krantz et al (1973)).

In practically all scenarios of interest to this project, dust layers are formed over impermeable surfaces, and the disturbing air is forced to flow parallel to the solid boundary. The no-slip flow condition at the boundary inevitably imposes a boundary layer type flow phenomenon around the dust layer. Figure 2.2 shows a boundary layer velocity profile typical for turbulent flow in the absence of dust entraiment. Figure 2.2 also shows the profile of root mean square streamwise velocity fluctuation which tends to be roughly 10% of the free stream velocity (Schlichting, 1968). As a rule of thumb, the peak of the root mean square transversal velocity fluctuation is roughly 5% of the free stream velocity, and occurs at a greater distance from the wall. The latter fluctuation component is a mechanism aiding migration of entrained particles away from the layer.

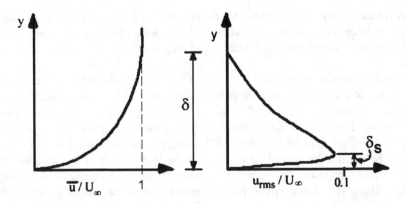

Fig. 2.2 Typical streamwise velocity profiles inside the boundary layer: mean velocity (left), root mean square of the fluctuating velocity (right)

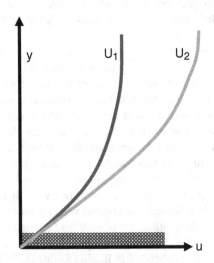

Fig. 2.3 Typical streamwise mean velocity profiles inside the boundary layer for two different friction coefficients and free stream velocities

The foregoing discussion suggests that aerodynamic forces acting on particles in a dust layer increase with increasing free-stream velocity. In early studies, the entrainment threshold or the entrainment amount had been correlated to the free stream velocity. In reality, small dust particles of interest here fall deep into the boundary layer as shown in Fig. 2.3. As a result, the layer is affected by the velocity gradient (or the boundary layer thickness) more so than the free stream velocity. That is why, in modern studies, the entrainment threshold or the entrainment amount had been correlated to the velocity gradient (γ), the shear stress (τ_w),

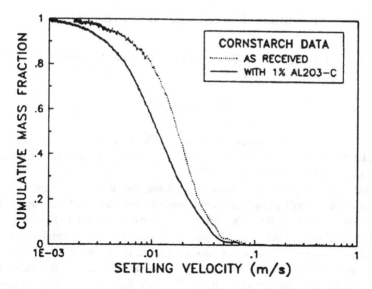

Fig. 2.4 Settling velocity of cornstarch with or without flowability agent Al2O3-C

or the friction velocity (interchangeably termed u_*, U_f or u_τ). By definition, these three parameters are uniquely related to each other through:

$$\tau_w = \mu\gamma = \rho u_\tau^2 \qquad (2.10)$$

It is also conventional to define a friction coefficient, C_{f_o} relating the friction velocity (u_τ) to the free stream velocity (U_e) through:

$$\tau_w = C_{f_o} 1/2\rho U_e^2 \qquad (2.11)$$

Where the subscript o designates the friction coefficient in the absence of dust entrainment.

The presence of a dust cloud in suspension complicates the turbulence field. The dispersed phase affects both the mean and fluctuating components of the flow field. A review of this subject can be found in Faeth (1987). As will be discussed later, the entrainment process also effects the boundary layer properties.

The scope of this discussion thus far is limited to spherical particles. Most real particles, however, are of non-spherical shape, which introduces another source of uncertainty to the analyses. Depending on the profiles they present, non-spherical particles can generate significant lift forces. Correction factors to the drag coefficient of some special cases of non-spherical particles have been developed; a review of this topic was given in Carmichael (1984).

Ural (1989) developed a test method to measure the settling velocity distribution of dust samples. Typical data is shown in Fig. 2.4 for corn starch with or without electrical conductivity additive.

2.3   Characterization of the forces Conserving the layer

For horizontal layers, gravity holds the particles in place. The weight of a spherical particle can be calculated from:

$$W = 1/6\pi\rho_p D^3 \qquad (2.12)$$

Other forces include adhesion/cohesion, chemical bonds and mechanical interlocking (as in long fibers). Frequently cited classical reviews of this topic include Corn (1966), and Zimon (1982).

Various methods have been used to measure the adhesive forces. When the adhesive force is smaller than the particle weight, the adhesive force can be determined from the tilting angle of the surface at the time particles fall. Similarly, in the pendulum method a particle is attached to a fiber and hangs freely. A surface (or another particle to measure cohesion) is placed in contact with the suspended particle, and then pulled away in a direction perpendicular to the suspending fiber, until the particle detaches from the surface. The adhesion force is calculated from the slope of the fiber at the moment of detachment. For larger adhesive forces, microbalance techniques are used where the particle is attached to a micro-force measurement system (which can be an electronic balance, a spring, or a cantilevered beam). Both the pendulum and microbalance techniques are limited to relatively large particle sizes because small particles are very difficult to attach to the suspending fibers.

Adhesion forces of smaller particles are measured more conveniently using the centrifuge and the vibration methods. Both methods are capable of measuring adhesive forces up to 10^6 times the particle weight.

The studies of adhesion have revealed that for particles of approximately identical size, under identical conditions, the forces of adhesion will not all be the same; in fact, they may span several orders of magnitude. Recognizing this highly statistical nature of the adhesion forces, an adhesion number is defined as the fraction of particles removed when subjected to a given force. The statistical nature of the adhesion/cohesion forces is also responsible for the kinetic behavior of the entrainment flux. In other words, under steady-state exposure to a turbulent aerodynamic disturbance, entrainment flux changes with time.

From the theoretical standpoint, a number of mechanisms are recognized as playing a role in adhesion/cohesion of powders in air. Molecules making up the particles possess attractive force fields (Van der Waals) around them, which in effect, hold them together. The force fields of the molecules near the particle boundary are not neutralized, thus providing an adhesion force. The force fields of the individual molecules have been integrated over the particle volume to obtain the net attraction force between two spherical particles of diameter D1 and D2. The result, called the Van der Waals Force, is

$$F_m \propto \frac{1}{Z_0^2} \cdot \frac{D_1 D_2}{D_1 + D_2} \qquad (2.13)$$

where, Z_0, the separation distance between the particles, is a major source of uncertainty[2] in determining the molecular forces. The attraction force between a spherical particle (of Diameter D) and a flat plate can be obtained by allowing D2 to go to infinity:

$$F_m \propto \frac{1}{Z_0^2} \cdot D \tag{2.14}$$

The constant of proportionality in these equations varies by orders of magnitude for different materials, and is generally higher for softer material (e.g., plastics versus abrasives) that can deform and provide an increased contact area. Presence of flaws and trace impurities are also known to affect the Van der Waals forces. The variation of. the Van der Waals forces under identical conditions is blamed (Zimon, 1982) on the "energetic inhomogeneity" of solid surfaces, as well as the microsurface roughness.

The second well studied mechanism of adhesion is due to capillary condensation (also known as formation of liquid bridges). The water vapor (or solvent vapors) may condense in the vicinity of contact of two bodies, even when the vapor phase is below saturation, because a negative curvature exists in the contact area, and the equilibrium vapor pressure is a function of surface tension and curvature. The condensed liquid forms a film that draws the two bodies together because of surface tension and capillary pressure. The diameter of the liquid bridge is usually small compared to the particle diameter so that the adhesive force due to capillary pressure is negligible compared to that for surface tension. For completely wetting smooth surfaces, the adhesive force between a spherical particle and a flat surface is given by

$$F_L = 2\pi\sigma D \tag{2.15}$$

where σ is the surface tension of liquid in contact with air, and D is the particle diameter. The adhesive force between two spheres of the same diameter is one half of the value calculated from this equation. Experimental evidence shows that the capillary condensation of water begins to occur at relative humidity levels above 70 percent.

There are two types of electrostatic forces that may play a role in particle adhesion. The first type arises from the contact potential, developed between the surfaces of two different materials. Ranade (1987) states that this type of force increases linearly with particle size, while Zimon (1982) recommends a two—thirds power dependence on particle size. The second type of electrostatic force is due to electric charges on the particles or the plate and is called the Coulomb force.

---

[2] The separation distance is usually not a directly measurable quantity for dusts, and assumptions for its value range typically from 0.4 to 1.0 nm.

For a spherical particle possessing an electrical charge, Q, resting on a flat uncharged surface, the Coulomb force is given by

$$F_c = \frac{Q^2}{6(Z_0 + D)^2} \qquad (2.16)$$

where is Z_0 the separation distance, and D is the particle diameter. This equation exhibits the reduction in Coulomb force with increasing particle size if the particle charge were to be constant. However, particle charge may also depend on the particle size. Dust particles dispersed by air demonstrate charges increasing slightly less rapidly than the square of the diameter, so that the Coulomb forces may also increase with the square of the diameter. It should be noted that the charge on particles contacting a surface will change with time due to electrical leakage. Another important feature of the Coulomb forces that is different from the other types of adhesion forces is that they decay relatively slowly with distance, and these forces may play a role even after the initial dislodgement (e.g., Owen (1969)).

Other types of adhesion mechanisms include magnetism, acid—base interactions (Ranade, 1987), capillary pressure in pore spaces filled with liquid, highly viscous binding agents, and crystal bridges (Rumpf, 1977).

It is clear from the foregoing discussion that the adhesion theory is far from being a predictive tool, at this time. The most important conclusion, however, is that adhesive forces are typically proportional to the particle size. Since the particle weight is proportional to the cube of the particle size, the forces holding the particle down in a layer is expected to be dominated by the adhesive force for the small particles, while it is controlled by the particle weight for large particles. This fact explains the reason why larger particles produce more repeatable and more predictable results.

For small particles, experimental studies sometimes produced contradictory results. Direct as well as inverse dependence of adhesive force on particle diameter (or sometimes even complete independence) has been reported. Zimon (1982) attributed those contradictions to the statistical nature of adhesive forces. The curves of adhesion number versus adhesion force for different sizes of the same material are usually not parallel to each other and curiously tend to cross each other. Depending on the location of the adhesion number taken to characterize adhesion with respect to cross-over part of the curves, contradictory results will be obtained.

2.4   Fundamental Studies of Particle Removal from Surfaces

This continues to be an active research area and a large body of experimental and theoretical work has already been published. A review of the topic can be found in Ziskind et al (1995) and Gradon (2009). Experiments indicate that adhesion forces as well as the aerodynamic forces exhibit a stochastic distribution. Coherent structures in the airflow play a significant role on the threshold entrainment conditions as well as entrainment rates. As a result, entrainment rate is not constant under specified conditions, but varies as a function of time.

Other experimental variables include underlying surface material, surface roughness, particle moisture, and the presence of an electrical field. In general, existing theoretical models are incapable of predicting the experimental data.

Some of the recent noteworthy publications include Ibrahim et al (2008), Jiang et al (2008), Merrison et al (2007), Grzybowski and Gradon (2005 and 2007), Masuda et al (1994), Gotoh and Masuda (1998), Hayden et al (2003), Rasmussen et al (2009), Roney and White (2006), Brateen et al (1990), Friess and Yadigaroglu (2002).

2.5 Applied Research on Aerodynamic Entrainment Threshold

At this point, it should be clear that first principle modeling of aerodynamic entrainment threshold or entrainment flux does not promise much success due to large gaps in current capabilities to predict the aerodynamic forces as well as the forces conserving the layer. For that reason, many studies have been carried out in wind tunnels or in open atmosphere and correlations have been proposed. While these correlations might be extrapolated to similar flow conditions and dusts, the major difficulty with this approach is that the results can not be generalized to all dusts.

Early work of Bagnold (1941) is still the most cited reference of the field. Bagnold studied the conditions leading to the saltation phenomenon by spreading a thick layer of sand on the bottom of a 30×30 cm cross-section wind tunnel using mean air velocities up to 10 m/s. The sand particles were typically 100 microns or more in size so that adhesive forces were negligible compared to particle weight. Bagnold determined that in order to initiate grain movement, the condition:

$$\tau_w > 0.01 \rho_p gD \qquad (2.17)$$

must be satisfied. In this equation, τ_w denotes the wall shear stress[3], whereas ρp, D, and g represent the particle density and diameter and the gravitational acceleration, respectively. Bagnold also discovered that the saltation, once initiated artificially at shear stresses below the value of initiation value, can sustain itself so long as:

$$\tau_w > 0.0064 \rho_p gD \qquad (2.18)$$

---

[3] In the literature, three parameters are used commonly to characterize the flow conditions near the wall: wall shear stress, τ_w, velocity gradient, γ; and friction velocity, u_τ (or u*). These three parameters are uniquely related to each other through the following relationship: $\tau_w = \mu \gamma = \rho u_\tau^2 = \rho u_*^2$

is satisfied. These equations are often referred to as the static and dynamic thresholds of saltation, respectively. The reason for the dynamic threshold being lower than the static threshold was explained by the ejection of new particles from the layer upon impact of saltating particles. Bagnold observed that the saltating particles leave the surface vertically at a velocity comparable to the friction velocity of the boundary layer. As discussed earlier, the aerodynamic lift acting on the particle is too small to explain this behavior. Therefore, the particle ejection is believed to be due to impact of either rolling or saltating particles. On this basis, however, it is hard to rationalize the observation that a factor of from two to twenty-five more mass is being conveyed by jumping than by rolling for various sand and soils (Fuchs (1964)).

Bagnold's data obtained for sand particles typically larger than 100 um have shown that smaller particles are moved more easily than larger particles. Numerous experiments performed later with smaller particles have shown that this trend is reversed for fine particles, so that the plots of threshold shear stress as a function of particle size have a minimum. The particle size for minimum threshold shear stress (i. e., for optimum dispersion) has been found to vary with the type of powder tested and may also be dependent on the details of the experimental conditions. For example, Zimon (1982) quotes optimal particle sizes of 15-20 um for sylvite dust and 100—150 um for corrundum particles laying on a steel wall. Similarly, Allen (1970) quotes an optimal diameter of about 100 um for quartz density sand. The increase of aerodynamic wall shear stress required to move smaller particles is widely believed to be due to adhesive forces. The adhesive forces are typically proportional to the particle size, whereas the gravitational force is proportional to the third power of the particle size, so that the former should dominate the latter for sufficiently small particles. The aerodynamic forces, reviewed earlier in this section, are typically proportional to the square of the particle diameter. Therefore, apparent contradiction in large versus small particle trends are explainable through force-balance theories.

Figure 2.5 shows the force-balance theory predictions for spherical sand particles ($\rho p = 2500$ kg/m3) in air by Phillips 1980. According to his theory, for large particles, line DE represents the condition

$$\text{Aerodynamic drag force} = \text{Particle weight}$$

which resulted in $\tau_w \alpha\, D$.

For small particles, line XY represents the condition:

$$\text{Aerodynamic lift} = \text{Particle adhesion force}$$

which resulted in $\tau_w \alpha\, D^{-4/3}$.

For intermediate size particles, Phillips postulated

$$\text{Aerodynamic lift} = \text{Particle weight}$$

which resulted in threshold shear stress being independent of particle size.

Numerous studies describing force-balance models and resulting threshold shear stress (or equivalently threshold friction velocity as defined in footnote 5) curves have been published.

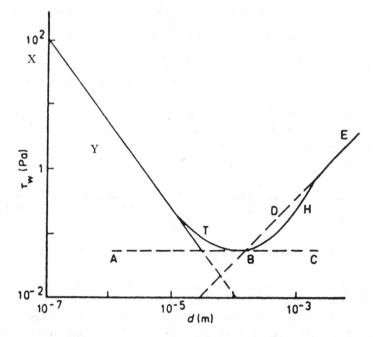

Fig. 2.5 Prediction of a force-balance theory for spherical sand particles in air

Figure 2.6 shows such a curve for threshold shear velocity versus particle size derived for wind erosion of sand (Greely and Iverson, 1985). Also superimposed on this figure are lines for constant value of the terminal velocity to friction velocity ratio delineating different modes of entrained particle motion.

For cohesive dusts, the threshold shear stress (or friction velocity) for particle movement is recognized to depend strongly on the conditions the deposit has been subjected to since its formation. For example, for erosion of desert soils, Gilette (1978) recommended actual field measurements using a portable wind tunnel with an open-floored test section.

A review of the Soviet activity on aerodynamic removal of powders from solid surfaces is given in Chapter 10 of Zimon (1982). The various experiments described by Zimon include removal of dust particles by air flow inside long ducts, detachment by a developing flow over a flat plate at various angles of attack, as well as detachment of particles from cylindrical surfaces by external air flow. Zimon at times has omitted some of the essential information in his review so that original papers may have to be referred to before using the data. In an interesting experiment described by Zimon (1982), spherical glass particles of varying diameter were placed on a steel plate with a Class IV surface finish[4] and were subjected to

---

[4]This is a Russian designation of surface roughness which corresponds to asperity height of 40 microns (1600 micro-inch).

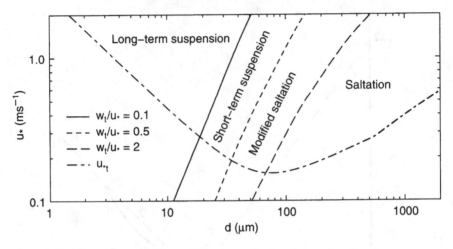

Fig. 2.6 Threshold shear velocity and suspension modes for wind erosion of sand

a 6.2 m/s free stream velocity. The shear stress on a flat plate decreases with increasing distance from the leading edge; therefore, close to the leading edge, all the particles are removed, but away from the leading edge none of the particles are removed. Zimon (1982) used the observed distances for complete removal and no removal to estimate the aerodynamic force required to remove the particles. There are several drawbacks associated with this type of experiment:

(1) the fact that the boundary layer equations are singular near the leading edge introduces some uncertainty in calculated flow conditions in this area;

(2) sufficiently away from the leading edge, the boundary layer may display transition to turbulence which introduces additional uncertainty to the calculated wall shear stresses;

(3) the removed particles transported downstream towards the undisturbed area may have an effect on the critical distance measured for no particle removal; and

(4) the shear stress, in the laminar flow regime, is inversely proportional to the square root of the distance from the leading edge, so that a relatively long plate must be used to observe both distances for complete and no particle removal in a single experiment.

Figure 2.7 shows the "lift-off" apparatus developed by Ural (1989) to determine the aerodynamic forces required to remove a dust deposit. Air moves radially inward between the two disks accelerating towards the center due to the reduction in the effective cross-sectional area. In addition to radial variation, the local air velocity is controlled by means of changing the gap between the plates and by adjusting the total airflow rate through the system. Calibration charts were developed to look up shear stress corresponding to a given radius (of the particle free circle created by the flow) at a given gap and pressure drop. Figure 2.8 shows a typical dust removal pattern which would be considered acceptable for the test method.

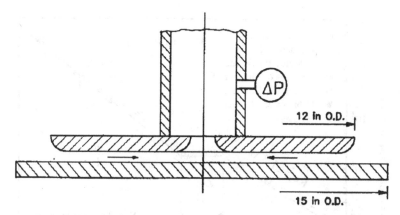

Fig. 2.7 Lift-off apparatus to determine the aerodynamic forces required to remove a dust deposit

Fig. 2.8 Dust removal pattern in the lift-off apparatus

Recently, Kalman et al (2005) proposed a general correlation based on particle Reynolds number, Rep, and the Archimedes number, Ar. The correlation, in the form of three piecewise continuous equations appear to correlate bulk velocity for pick up threshold, U_{pu}, in gases as well as in liquids:

$$\text{Zone I} \quad Re_p^* = 5Ar^{3/7} \tag{2.19}$$

$$\text{Zone II} \quad Re_p^* = 16.7 \tag{2.20}$$

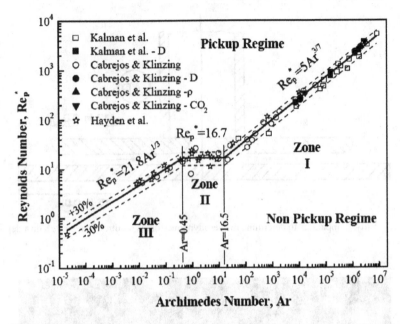

Fig. 2.9 Comparison of Kalman et al (2005) correlation with data

$$\text{Zone III} \quad Re_p^* = 21.8Ar^{*1/3} \tag{2.21}$$

Where:

$$Re_p^* = \frac{\rho \cdot U_{pu} \cdot d}{\mu \cdot \left(1.4 - 0.8 \cdot e^{-\frac{D/D_{50}}{1.5}}\right)} \tag{2.22}$$

$$Ar = \frac{g \cdot \rho \cdot (\rho_p - \rho) \cdot d^3}{\mu^2} \tag{2.23}$$

d: particle diameter
D: pipe diameter
μ: dynamic viscosity
g: gravitational acceleration
ρ: fluid density
ρp: particle density
$Ar^* = 0.03 \, e3.5\phi \, Ar$, modified Archimedes number
ϕ: particle sphericity

A correction factor for non-spherical, relatively large particles was also provided. Figure 2.9 shows the goodness of how their correlations match the gas pick-up data,

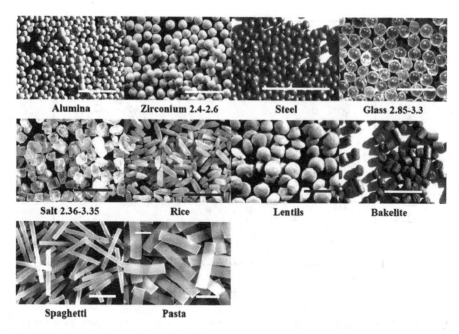

Fig. 2.10   Shapes of non-spherical particles

and Fig. 2.10 shows types of particles used to validate the non-spherical particle correction factor. Fig. 2.11 shows the threshold (pickup) velocities calculated for spherical glass particles.

2.6   Applied Research on Aerodynamic Entrainment Mass Flux

The measurement of this parameter is significantly more difficult than the entrainment threshold.

Measured values of dust emission by wind erosion typically range from 10^{-7} to 0.1 g/m2-s, because of the relatively low velocities experienced in natural winds. Gillette (1977) carried out indirect measurements of entrainment mass flux for nine different soils as a function of friction velocity. The large degree of scatter in the data did not permit development of a definitive correlation. Data seem to support the power relationship between the mass flux Φ and shear velocity u_* through

$$\Phi \alpha u_*^n \tag{2.24}$$

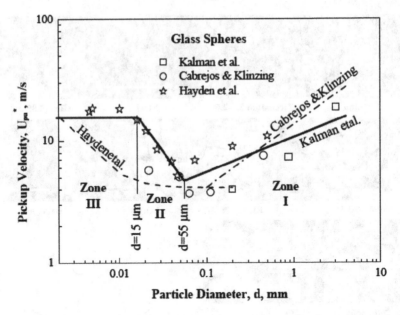

Fig. 2.11 Comparison of Kalman et al (2005) correlation with data for glass spheres and other correlations

Where exponent n could range from 3 to 5. Later, Gillette and Passi (1988) suggested a relationship, which takes threshold entrainment velocity, u_{*t} into account:

$$\Phi = \alpha_g u_*^4 [1 - u_{*t}/u_*] \qquad (2.25)$$

where α_g is a dimensional coefficient. It should be noted that, at high free stream velocities, boundary layer is affected by the large entrainment mass flux and the exponent n could be as low as unity.

Hartenbaum (1971) performed a limited number of steady state entrainment tests in a wind tunnel which had a test section 40″ high and 18″ wide. Free stream velocities ranged from 34 to 115 m/s. A particulate bed approximately 4″ deep at the start of each test was composed of AFS 50-70 Ottawa silica testing sand which had a mean particle diameter of 250 um. The entrainment rate was correlated to the free stream velocity U with the following equation:

$$\Phi(1b/ft2 - s) = 0.366 \; 10^{-2}[U(ft/s)]^{5/4} \qquad (2.26)$$

Hartenbaum also took time to characterize the boundary layer at the test section and fitted a correlation to the shear velocity:

$$\Phi(1b/ft2 - s) = 0.86 \; 10^{-3}[U_*(ft/s)]^{5/4} - 0.01 \qquad (2.27)$$

Later, upon request from the US Bureau of Mines, Rosenblatt recast the Hartenbaum's free stream velocity equation to include air density effect in an ad hoc fashion. He also included an ad hoc threshold free stream velocity effect to make the entrainment flux nil at 420 cm/s for mine applications. The resulting equation:

$$\dot{m}'' = \rho \cdot U \cdot \left[0.0021 \cdot U^{0.25} - \frac{4}{U} \right] \tag{2.28}$$

is still being used by NIOSH for mine research purposes (Edwards and Ford, 1988). Bureau of Mines measurements (Singer, Harris and Grumer, 1976) indicate that gas-explosion induced air flow threshold velocities are in the range of 5 to 30 m/s for coal dust. It is noteworthy to point out that both Hartenbaum and Bureau of Mines correlations, mass flux is proportional to the 1.25 power of the free stream velocity.

Batt et al. (1995) and Batt et al (1999) reported extensive entrainment data for high-speed air flow velocities, typically ranging from 100 to 300 ft/s. They developed the following correlation:

$$\dot{m}'' = (0.3 \pm 0.1) \frac{\rho_e \cdot U_f \cdot M_e^{0.5}}{\alpha / \alpha_{WMSR}} \tag{2.29}$$

where:

m'': entrained mass flux
ρ_e: free stream air density
U_f: friction velocity
M_e: free stream flow Mach number
α: angle of repose of soil
α_{WMSR}: angle of repose for soil at the White Sands Missile Range

This equation appears to correlate well the entrainment rate of Ottawa Sand and White Sands Missile Range soil, over a wide range of parameters tested, and predicts mass flux to be proportional to the 1.5 power of the free stream velocity. Batt et al (1995) point out that the larger exponent of 3, proposed originally by Bagnold in 1941, may more appropriate for wind erosion where the free stream velocity is typically below 20 m/s.

The Batt equation above expresses the mass flux as a function of the friction velocity, U_f. For the ease of use, it may be preferable to recast equation on the free stream velocity, even at a cost of precision loss. Since the dust entrainment occurs deep in the boundary layer, friction velocity rather than the free stream velocity is the more appropriate parameter to correlate the entrainment rate. On the other hand, most users of the NFPA standards are not anticipated to be versatile in using aerodynamics concepts encompassing the friction velocity. Therefore, an additional simplification is introduced by translating the selected entrainment rate correlation to free stream velocity. The original (2.29) is further modified here for low flow velocities so that entrainment rate tends to zero at the threshold velocity. Thus, the new equation proposed in this project is:

$$\dot{m}'' = (0.002) \cdot \rho_e \cdot U_e \left[U_e^{1/2} - \frac{U_t^2}{U_e^{3/2}} \right] \tag{2.30}$$

Where, U_e and U_t respectively represent the free stream velocity and the threshold (pickup) velocity. Equation (2.30), which constitutes the basis of our new strawman method described in the next chapter, is an improved version of the NIOSH equation (2.28), and includes an ad hoc correction for the appropriate threshold velocity. The coefficient 0.002 of (2.30) was selected to envelope the Batt et al data.

Zydak and Klemens (2007) studied the entrainment rates of dust deposits by airflow. Tests were conducted in a shock tube / wind tunnel with internal cross section 72 mm by 112 mm. Dust layer dust thickness tested were 0.1 mm, 0.4 mm or 0.8 mm. The following dimensional correlation was proposed by these authors:

$$\Phi = 0.004 h_1^{0.216} \; U^{1.743} \; D^{-0.054} \; \rho_p^{-0.159} \; Ap^{0.957} \tag{2.31}$$

Where:

Φ: entrained mass flux in kg/m2-s
h_1: layer thickness in mm
U: flow velocity above the layer in m/s
D: characteristic particle size um
ρ_p: particle density in kg/m3, and
Ap: is a dimensional empirical constant.

This is in fact the dust entrainment correlation built into the current version of the DESC code. Equation 2.31 assumes that the entrainment mass flux is proportional to the 1.75 power of the free stream velocity.

The authors selected the following input parameters when developing their correlation:

Coal dust	1.2		1340
Potato starch	0.745	75	1469
Potato starch	0.7	35	1527
Silicon dust	1.037		2341

Shock tubes have been recognized as a valuable tool in studying aerodynamic dust lift—off because they provide a well defined flow environment. They also have a direct application in characterizing the dust lift-off by blast waves emanating from conventional or thermonuclear explosions. Some examples of this type of experiment performed with non- cohesive dust are given by Gerrard (1963), Fletcher (1976),

Boiko et al (1984), and the references therein. In interpreting shock tube data, one must be careful about the effect of streamwise compression of dust layer across the shock wave. This effect often causes a significant lateral dust ejection velocity and throws the particles beyond the viscous boundary layer.

Fletcher (1976) has subjected layers of treated (free flowing) limestone dust, typically 14 um in size, to Mach 1.15 to 1.3 incident shock waves. The convective flow velocities under these conditions vary between 80 and 150 m/s. The dust cloud shapes measured from photographs were shown to fit the ballistic trajectories of individual particles having an initial vertical velocity of about 14.5 m/s. Boiko et al (1984) have repeated Fletcher's experiments with a much stronger shock (Mach 2.6 corresponding to a convective velocity of 628 m/s). They report vertical ejection velocities of 40 m/s for 200 um glass particles ($\rho_p = 1200$ Kg/m3), and 17 m/s for 200 um bronze particles ($\rho_p = 8700$ Kg/m3).

The small cohesive particles may be lifted from the surface in the form of aggregates. The breakdown of these aggregates in a turbulent flow field is of importance in determining the extent of the dispersion. Singer et al (1976) noted mine dust deposits that have been wetted or have undergone a wetting-drying cycle may constitute a greater explosion hazard than untreated dusts owing to selective lifting of relatively large briquetted fragments which then dispersed in the air stream.

The theoretical progress in the field has been at a rather slow pace. A good physical description of the various phenomena encountered in the pneumatic transport of non-cohesive powders is given by Owen (1969).

Corn and Stein (1965) used the calculated particle drag force to interpret their aerodynamic dust entrainment threshold data. The drag force was calculated using the spherical particle drag coefficient, and the boundary layer velocity at one particle radius distance from the wall. For less than 53 um diameter glass beads deposited on glass slide, these authors report that at removal efficiencies exceeding 75%, the calculated air drag was Within a factor of 2.5 of the adhesion force measured using the ultracentrifuge method. At lower removal efficiencies, the two forces differed by as much as a factor of 10. Corn and Stein also report that for the turbulent boundary layers employed in their experiments, the dust removal efficiency is somewhat dependent on the test duration. Later, Zimon (1982) repeated these experiments using 20 and 35 um loess[5] particles and reported good agreement with the centrifuge method. The formulas used by Corn and Stein are at best rough estimates of the drag force (parallel to the surface) exerted by air as it is calculated from the undisturbed velocity in the boundary layer at the level of particle center. As was stated earlier, the actual drag force for creeping flow was 70 percent higher than that calculated by the Corn and Stein method. Furthermore, in turbulent flow, some form of a peak force (rather than mean) should be responsible for particle dislodgement.

---

[5] Loess is an unstratified, usually buff to yellowish brown, loamy deposit found in North America, Europe, and Asia and is believed to be chiefly deposited by the wind.

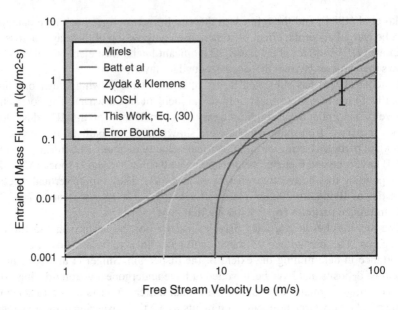

Fig. 2.12 Comparison of the predictions from various entrainment mass flux equations for 53 micron coal dust

Another interesting hypothesis advanced in Zimon's book was that the maximum diameter of the adherent particles remaining on a surface after being exposed to a turbulent boundary layer is equal to the thickness of the laminar sublayer. Zimon suggests the use of this hypothesis as a means to determine the laminar sublayer thickness. This hypothesis, although demonstrated with some experiments, is not completely substantiated.

The model of saltation phenomena given by Owen (1964) avoids these fundamental questions on particle removal using two key phenomenological hypothesis: (1) the effect of the moving grains on the fluid outside the saltation region is similar to that of solid roughness of height comparable with the depth of the saltation layer, and (2) the concentration of particles within the saltation layer is maintained to keep the wall shear stress at the dynamic threshold of Bagnold, given in (2.18). These two assumptions are validated by the agreement between the measured and predicted quantities. Owen (1964) also speculated that the particles would enter into suspension when the wall shear stress is of the same order of magnitude as $\rho_p g D$.

Recently, Mirels (1984) has used Owen's second hypothesis to calculate the dust erosion rates in developing turbulent boundary layers over a flat plate and behind a shock wave. Treating the effect of dust entrainment on the flow with his previous transpiration model, Mirels reported agreement Hartenbaum's high entrainment rate with data within a factor of two. It is remarkable, if not fortuitous, that Mirel's simple conceptual model predicted the experimental velocity exponent very well.

Figure 2.12 compares the predictions of the several entrainment mass flux equations described in this section. Calculations were made for 53 micron coal dust particles with a particle density of 1.34 g/cm3. Air density is taken to be 1.2 g/3. Kalman et al (2005) correlation results in a threshold (pick-up) velocity of 9 m/s for the onset of entrainment. The critical shear stress needed for the Mirels model was estimated to be 0.15 Pa from the threshold velocity, (2.10) and (11) using a friction coefficient of 0.003. The layer thickness needed for the Zydak and Klemens equation was assumed to be 1/32″.

All equations seem to agree with each other within a factor of three for the high speed flows. The disparity for low speeds were not unexpected since the NIOSH equation was forced to predict zero entrainment at 4.2 m/s, whereas Batt et al, and Zydak and Klemens correlation tend to no entrainment only when the free stream velocity approaches zero.

2.7 Secondary Explosion Propagation Tests

In these experiments combustible dust deposited on the floor of a gallery is dispersed and ignited by a primary explosion at the beginning of the gallery. Measurements typically include pressure development, apparent flame speeds, and gas velocities. Examples of these types of experiments can be found in Tamanini (1983), Richmond and Liebman (1974), and Michelis et al (1987). These experiments are quite costly, and therefore have been performed using very few types of dusts (mostly coal dust, and some cornstarch). Since a number of phenomena play crucial roles in sequence in these experiments, the results are often not repeatable. In order to improve the repeatability problems, the tests are designed so that the secondary explosion is overdriven by a strong primary explosion.

Similar tests have been carried out at intermediate scale. Tamanini (1983) used a 6 m long model gallery with 0.3 m2 cross-section to study secondary explosions of cornstarch. Recently, Srinath et al (1987) tested a number of dusts in their 0.3 m I.D. 37 m long flame acceleration tube. The scaling of test results from intermediate scale to actual size galleries should be difficult, at best.

Investigations aimed at understanding the aerodynamic dust entrainment in mine galleries have been carried out at the U.S. Bureau of Mines, as an extension of the early British work (Dawes, 1952).

Singer et al (1969) measured the minimum air velocities required for dispersal of coal and rock dust deposited at the floor of a small wind tunnel test section (7.62 cm wide, 2.54 to 5.08 cm high). Tests have been carried out using monolayer dust deposits, as well as piles of dust. The effects of type of surface holding the dust, relative humidity of the dust pre- conditioning atmosphere, and the large-amplitude oscillations superimposed on the air stream on the threshold dust entrainment velocity were studied. The air flow rate was transient during the tests, with a reported rise time of 1 minute.

In the monolayer studies, the average air velocities required to remove 25, 50 and 75 percent of the particles (by microscopic number count) were determined.

The measured threshold velocities for 75% dust removal ranged from 20 to 130 m/s, increasing with decreasing particle size. The calculated wall shear stresses in this configuration range from 1 to 30 Pa. The measured threshold entrainment velocities for the three types of dust tested increased in the order: rock dust, anthracite, and Pittsburgh seam coal. The differences in the threshold velocities of these dusts diminished for particle sizes below 10 um. The rock dust was removed more easily from smooth Pittsburgh seam coal and glass surfaces than from smooth anthracite.

In tests with dust piles, the minimum air velocity required for complete removal of the dust pile was measured. Velocity measurements were taken upstream of the pile, at a distance from the tunnel floor equal to mid-height of the ridge. The types of dust removal observed included erosion, denudation, as well as removal of massive clumps and sliding of the entire ridge. The reported threshold velocities spanned the range from 5 to 23 m/s. It was found that the compaction of ridge significantly increases the threshold entrainment velocity, whereas the presence of vibrations either in the air flow or on the floor reduces it. The relative humidity of dust in the range of 35 to 90 percent was found to have no significant effect on the threshold velocity. Interestingly, for the dust piles tested, anthracite dust was easiest to be removed, while rock dust was the most difficult, an order different than observed for monolayers.

Singer et al (1972) have later attempted to relate the entrainment threshold of dust piles to the shear cell data believed to be some representation of the cohesive forces between the particles. First, they have defined a threshold Froude number:

$$Fr = \frac{\tau_w}{\sqrt{\rho_b g \tau_y H / 2}}$$

where:

τ_w aerodynamic wall shear stress;
τ_y shear cell yield stress extrapolated to no load;
ρ_b bulk density of powder;
gravitational acceleration; and
height of the dust piles.

The denominator was stated to be the "geometric mean of the gravitational and cohesive forces," yet it lacks any physical significance. Singer et al (1972) found that for their limited number of data points, this Froude number remained relatively constant within the range 0.0077 to 0.038 for their dust ridges and beds. These authors also tried the ratio of the aerodynamic dynamic pressure at the mid-height of the pile to Ty and found it to cover the range between 0.22 and 0.75.

A limited number of threshold tests were repeated in a large scale (1.5 m high, and 2.4 m wide) wind tunnel, which indicated that the threshold velocities in the large scale wind tunnel is a factor of 1 to 3 smaller than those measured in the small wind tunnel. These authors have also made some entrainment rate measurements and presented their results as data correlations. These correlations must be used with extreme caution beyond their intended range or for different dusts because they are not based on physical reasoning.

In a follow-up, work Singer et al (1976) have used actual explosion induced air flow, fraction of a second in duration, to study the dust dispersion phenomena. Most tests were carried out in a 0.61 m I.D. 49.7 m long explosion tunnel, while some tests were repeated in the full scale experimental mine gallery. The instantaneous threshold air velocities in these explosion tests were found to be in the same range as the earlier slow-rise threshold velocity tests described above. One of the important objectives of this study was to study the selective dispersion of coal dust over rock dust, which would decrease the inerting probability. It was found that the uniformly mixed beds always dispersed without separation, whereas in the case of coal dust layer deposited over a rock dust layer, only the coal dust is dispersed if the peak airflow velocity is in a range between the threshold velocities of the two dusts.

Hwang et al (1974) modeled the dispersion phenomena using the diffusion equation. The details of the entrainment were completely ignored and the entrainment rate, specified as a denudation rate, was left as input. This model also ignores the effects of gravity. There is also a great uncertainty in picking a diffusion coefficient which was stated to be between 0.2 to 362 cm2/s. These authors recommended the use of diffusion coefficients in the range 25 to 100 cm2/s as "best guess," which results in a factor of four variation in the calculated dust concentration.

The modeling effort at the University of Michigan was focused on the coupling of the dust lift-off with flame propagation. In this model, the one-dimensional flame propagation model of Chi and Perlee (1974) was mated with the entrainment rate model of Mirel (1984) described above. The agreement with the data given in Srinath et al (1987) was qualitative. The limitations of this approach arose from the fact that the mixing process and the effects of gravity were not included. More recently, Li et al (2005) examined the possibility of deflagration to detonation transition supported by layers of corn dust, cornstarch, Mira Gel starch, wheat dust, and wood flour. Flame speeds of up to 1300 m/s were observed, which the authors called quasi-detonations.

2.8 Computational Simulation of Aerodynamic Dust Entrainment Phenomena

Computational Fluid Dynamics (CFD) tools have also been used to predict the entrainment phenomena. Iimura et al (2009) studied the removal agglomerates by shear flow, using a modified discrete element method. Ilea et al, in University of Bergen developed an Eulerian-Lagrangian model and studied various aspects of dust entrainment behind a shock wave.

Dust Explosion Simulation Code (DESC) is a CFD models sometimes used to simulate secondary explosions. However, as was discussed above, these models rely on very crude correlations to represent the entrained mass flux. Hence, it is hoped that the CFD models too will benefit from this project.

2.9 Gaps in Available Information

This study revealed serious gaps in the available information.

Even though a large body of fundamental experimental and theoretical work has already been published, existing theoretical models are incapable of predicting the experimental data. This difficulty is inherent in the phenomena involved in dust entrainment, as experiments indicate that adhesion forces as well as the aerodynamic forces exhibit a stochastic distribution. Coherent structures in the airflow play a significant role on the threshold entrainment conditions as well as entrainment rates. As a result, entrainment rate is not constant under specified conditions, but varies as a function of time. Additionally, factors such as underlying surface material, surface roughness, particle moisture, and the presence of an electrical field also display significant effects.

Applied research focused on specific applications thus resulting in more encouraging predictive tools. There exist a significant number of publications, which can be distributed into clusters such as atmospheric erosion, pneumatic transport, fluidization, pharmaceutical delivery, atmospheric emission. Unfortunately, due to limitations in applicability, these studies are not directly useable for the present project.

There exist a limited number of experimental studies secondary explosions. However, scale, geometry and the parameters of these tests limit their generalization to the broad range of industrial applications.

Chapter 3
Proposed Strawman Method

3.1 An introduction to the Proposed Strawman Methodology

Recently, the Committee responsible for NFPA 654 came up with two new consensus criteria for determining that a compartment is a dust explosion hazard: one aimed at mitigation of burn injuries, and the other for room/building collapse prevention. As described in Rodgers and Ural (2010), the criteria are based on maximum allowable airborne combustible dust mass. Both formulas rely on an empirical entrainment fraction, η_D, representing the fraction of dust accumulations that can become airborne during an accident. After much discussion, the Committee selected a value of $\eta_D = 0.25$ which offers the same level of protection NFPA 654-2006 does for typical occupancies, pending the outcome of this Research Foundation sponsored project.

The objective of the method is to estimate the amount of dust that can be removed from dust layers by a primary event such as a pressure vessel burst or a primary explosion, thus eliminating the need to use an empirical value for entrainment coefficient. The prediction will obviously depend on the nature and the strength of the primary event, the air velocity it induces over the layer as a function of location and time, and the resistance of the dust in the layer against entrainment.

Mathematically, if a primary event is capable of inducing velocity u=u(x,y,t) over the layer, and the entrainment mass flux for the particular dust is given by the expression $m'' = m''(u)$, then the total mass of dust removed from the layer, M, can be expressed as:

$$M = \iiint m''[u(x, y, t)] \cdot dx \cdot dy \cdot dt \tag{3.1}$$

Since such a rigorous approach is impractical for the anticipated end users, a simplified approach was sought. The simplification was achieved by narrowing

E.A. Ural, *Towards Estimating Entrainment Fraction for Dust Layers*,
SpringerBriefs in Fire, DOI 10.1007/978-1-4614-3372-9_3,
© Fire Protection Research Foundation 2011

down the initiating events to a few typical primary event scenarios. Additional simplification was achieved by breaking the methodology into several components and further simplifying them. These components include:

- Estimation of threshold entrainment velocity for dust
- Estimation of entrained mass flux
- Estimation of the flow velocity and duration induced by the primary event
- Estimation of total entrained mass

While this approach can result in some loss of generality and precision, its ease of use is obviously a major benefit.

3.2 Estimation of Threshold Entrainment Velocity for Dust

Empirical algebraic relationships proposed by Kalman et al (2005) were selected for use in this phase of the project. Three expressions were provided for different particle size groups, accounting for particle size and particle density. An additional equation was provided to account for the particle shape for the large particle regime.

The end users of this methodology are expected to refer to a chart similar to that shown in Fig. 3.1, which was created using the Kalman et al equations. For example, if the particular dust were made up of atomized aluminum particles of 100-micron diameter, then no entrainment would be expected so long as the free stream velocity, U_t, over the layer is below 7.5 m/s, as read from Fig. 3.1.

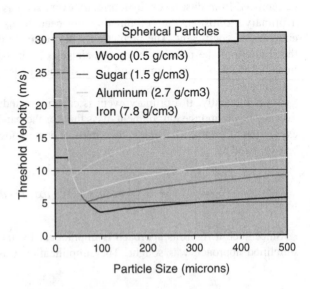

Fig. 3.1 Calculated threshold entrainment velocity as a function of particle size and particle density

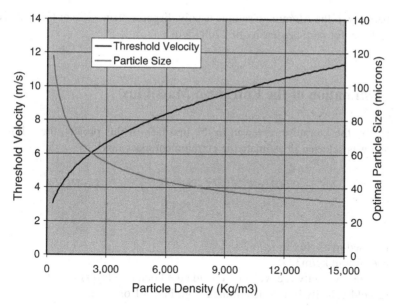

Fig. 3.2 Variation of the minimum threshold velocity and the corresponding optimal particle size with particle density for nearly spherical particles

Alternatively, the minimum value of the appropriate curve can be used for poly-disperse dusts (i.e. dusts made up of particles with a broad size range). For nearly spherical particles, the minimum threshold velocity (in m/s) is:

$$U_t = 0.46\rho_p^{1/3} \tag{3.2}$$

and it corresponds to the optimal particle size (in meters):

$$D_{opt} = 7.9^*10^{-4}\rho_p^{-1/3}$$

Where ρ_p is the particle density in kg/m3. Variation of the minimum threshold velocity and the corresponding optimal particle size with particle density is charted in Fig. 3.2. In applications where dust particles are expected to be removed as agglomerates, substituting particle density with bulk density may be more appropriate. Non-spherical particle shapes are treated using a correction factor based on the particle sphericity.

The end users of this methodology are expected to refer to a chart similar to that shown in Fig. 2.1, which was created using the equations provided in Kalman et al (2005). For example, if the particular dust were made up of atomized aluminum particles of 100-micron diameter, then no entrainment would be expected so long as the free stream velocity, U_t, over the layer is below 7.5 m/s, as read from Fig. 3.1.

Alternatively, if the Aluminum sample is poly-disperse, then (3.2) or Fig. 3.2 gives $U_t = 6.4$ m/s for particle density of 2700 kg/m^3.

3.3   Estimation of the Entrained Mass Flux

Inspired by the literature reviewed in the first task of this project, the following equation was selected to estimate the entrainment mass flux[1]:

$$m'' = 0.002 * \rho * U * \left(U^{1/2} - U_t^2 / U^{3/2} \right) \qquad U > U_t \qquad (3.30)$$

where:

m'' entrained mass flux in kg/m2-s
ρ gas density in kg/m3
U free stream velocity (i.e. velocity outside the boundary layer) in m/s
U_t threshold velocity in m/s determined from Fig. 3.1 or (3.2).

This equation predicts that the entrainment mass flux is proportional to the 1.5 power of the free stream velocity and goes to zero when the free stream velocity approaches the threshold velocity.

3.4   Comparisons with Large Scale Explosion Data

The Equations (3.30) and (3.2) represent the essence of the new strawman methodology. Their predictions are compared to the available test data in this section.

NIOSH TESTS: The National Institute for Occupational Safety and Health (NIOSH), Office of Mine Safety and Health Research (OMSHR) has conducted large-scale dust explosion tests in an experimental mine (Cashdollar et. al, 2010). The gallery is approximately 1600 ft long, 7 ft high and 20 ft wide. Before each test, the gallery was thoroughly washed down. Dehumidified air was passed through the gallery, and the gallery was allowed to dry several days before dust was loaded. As seen in Fig. 3.3, tests were driven by the ignition of a methane air mixture. Typically, the first 40-ft (12 m) section of the mine gallery, starting at the face (closed end) was filled with 10% methane in air mixture. A plastic diaphragm was used to contain the methane-air mixture within the flammable gas mixture zone before ignition. The coal dust and limestone rock dust mixture was placed half on roof shelves made of expanded polystyrene and half on the floor as illustrated in Fig. 3.3.

---

[1] The rate of mass removal per unit area per unit time.

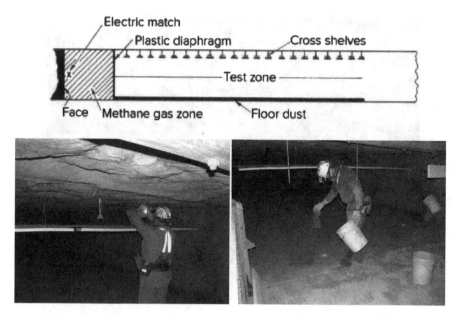

Fig. 3.3 Experimental setup in NIOSH tests. Top: Side view schematics of the mine geometry, Left: Placing Coal/Rock dust mixtures on the shelves, Right: distributing dust on the floor (from Cashdollar et al, 2010)

These roof shelves were suspended 1.5 ft (0.5 m) from the mine roof on 10-ft (3 m) increments throughout the dust zone which was 300-ft long (i.e. spanned the 40 ft to 340 ft distance from the face. The roof shelves were often damaged during the deflagration, and resulting debris occasionally landed on the test bed causing easily identifiable gouges on its surface. The amount of the dust mixture placed in the dust zone corresponds to a nominal dust loading[2] of 200 g/m3. Ignition of the methane-air mixture alone would result in flame travel up to approximately 200 ft from the closed end. The methane air zone was ignited near the center of the face using electric matches grouped in a single-point configuration.

Based on personal communications with Marcia Harris, a Research Engineer with OMSHR, the dust removal experiments within the NIOSH experimental mine were setup and conducted in the following manner (Harris et. al, 2009). The dust bed for the dust removal tests (Fig. 3.4) was prepared at a location 250 ft (to 258 ft) from the face (ignition end) between two aluminum rails. These rails were 1-inch high by 100-inches long, parallel to the gallery axis, and attached to the mine floor 22-inches apart. The dust was placed between the rails and leveled; creating a 1-inch deep layer. Before and after the explosion test, the dust layer depth

---

[2] The nominal dust loading assumes that all of the dust was dispersed uniformly throughout the cross-section.

Fig. 3.4 Measuring the amount of dust scoured during an explosion (from Harris et al, 2009)

was measured, to within ± 0.1 mm accuracy, at stations 24″, 36″, 48″, 60″, 72″, 84″ and 96″. The depth change at each station was attributed to the dust removal due to explosion. Time resolved gas flow velocity induced by the primary explosion near the dust bed was also recorded using a bi-directional velocity probe located few feet downstream of the bed near the center of the mine cross-section.

Figure 3.5 shows the velocity recorded near the dust bed (containing a 35% coal dust 65% rock dust mixture) in experimental mine test 511. The dust removal depth shown in this figure was calculated from the new strawman method. Equation (3.2) estimates the dust bed threshold velocity, Ut, to be 6.4 m/s[3]. Inserting the instantaneous gas velocity and gas density into Equation (30), dust removal rate (entrainment mass flux) is calculated as a function of time. Cumulative mass removal per unit area (kg/m2) is calculated as a function of time by integrating entrainment mass flux over time. Dust bed bulk density of 850 kg/m3 was used to convert the cumulative mass removal to dust removal depth shown in Fig. 3.5. Our strawman methodology is seen to predict that the particular explosion created in this test will remove approximately top 2.4 mm of the 25 mm thick dust bed, corresponding to a 9% entrainment fraction. Dust removal observed in the test ranged from 1 mm to 2.5 mm depending on the location along the bed (Fig. 2.6).

---

[3] The particle densities for the coal dust and rock dust are 1330 and 2750 kg/m3, respectively. Since tested mixtures were mostly rock dust (64 to 80% rock dust by weight), rock dust particle density was used to estimate the threshold velocity.

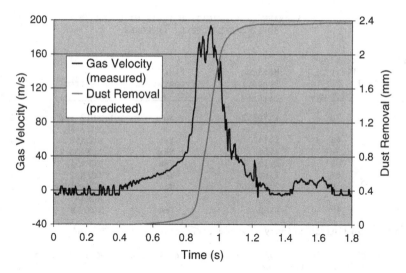

Fig. 3.5 Gas velocity recorded near the dust bed during the experimental mine test 511, and the corresponding dust removal depth calculated from Equation (30)

Figure 3.5 shows the experimental dust removal depths determined from pre and post explosion depth measurements (Harris, 2010). Table 3.1 lists the key results obtained in all NIOSH tests conducted in this particular program. Measured dust removal depths reported in the sixth column represents the average of the depth changes observed in stations 36″, 48″, 60″, 72″, and 84″. The removal data from the first and last stations were ignored due to potential edge effects. The last column of Table 3.1 shows the predictions of the new Strawman method, calculated from the velocity data as described above. The strawman methodology is seen to provide a reasonable representation of the data.

B. NGFA TESTS: Another large-scale explosion study, where the quantity of dust removal was carefully measured, was reported by Tamanini (1983). The tests were conducted in an 8′ by 8′ by 80 feet long gallery connected at one end to a 2250 ft3 chamber. The other end of the gallery was open to atmosphere. Before each test, a uniform layer of cornstarch was laid on the gallery floor using a modified lawn spreader. After each test, the gallery was swept and the dust residue was collected and weighed. Primary explosion was created by igniting a cornstarch cloud (nominally 125 g/m3) formed in the primary chamber. The table below, taken from Tamanini 1983, lists the tests performed, the amount of cornstarch spread over the gallery floor (first column), and the amount of cornstarch picked up (second column) by the explosion, reported as a corresponding average concentration in the gallery.

SUMMARY OF PROPAGATION TESTS WITH THE
80-FT GALLERY FULLY ENCLOSED

Fuel Concentration* in the Gallery [g/m³]		Maximum Pressure [psig]	Flame Propagation to 80 ft	Data from Test #	Notes
Nominal	Pickup				
0	0	1.50+.20	No	FMDU17 and NGFA49	1
77	48	2.14	Yes	NGFA50	1
105	36	1.83	Yes	NGFA53	1
155	62	2.13	Yes	NGFA51	1
285	86	1.48	Yes	NGFA52	1
0	0	2.55+.55	No	NGFA54 and NGFA55	2
93	49	2.25	Yes	NGFA56	2

*Dust loading in primary chamber corresponding to 125 g/m³.

1. Dust charges for the primary chamber in the middle of 10-ft long shelves at 6.7 ft above the floor

2. Dust charges for the primary chamber in 5-ft long shelves placed on the floor.

Chart A

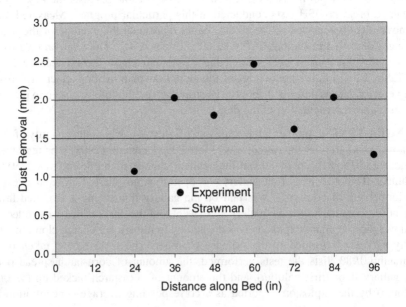

Fig. 3.6 Experimental mine and theoretical dust removal depths for test 511

Table 3.1 Key NIOSH Results

Test No.-Test Bed Location	% Rock Dust in Coal Dust	Coal Dust type in Mixture	Pressure Data		Dust Removal	
			Dynamic Pressure Peak	Dynamic Pressure Impulse[4]	Measured	Strawman Method Prediction
	%		psi	psi-sec	mm	mm
498-125ft	80	PPC	18.9	0.4	1.0	1.1
498-320ft	80	PPC	3.3	0.3	1.2	1.0
499-125ft	65	PPC	2.6	0.4	2.1	1.1
499-227ft	65	PPC	5.4	0.5	1.6	1.5
499-320ft	65	PPC	9.7	0.2	2.1	1.5
511-250ft	65	PPC	7.1	0.9	2.0	2.4
512-250ft	75	PPC	1.3	0.3	2.6	1.3
513-250ft	80	PPC	1.1	0.3	1.0	1.0
514-250ft	64	Coarse	6.4	0.8	2.1	2.1
516-250ft	69	Coarse	4.8	0.6	1.2	1.8
517-250ft	71.7	Medium	3.8	0.5	1.6	1.6
518-250ft	74.4	Medium	4.2	0.5	2.1	1.5
520-250ft	68.5	Medium	6.0	0.7	1.4	2.2

Notes: PPC denotes Pittsburgh Pulverized Coal that contains 80% fines that go through 200 mesh (74 microns). Medium and Coarse Coal Dust respectively contain 40 and 20% fines that go through 200 mesh. The tests 498 and 499 each employed a 70 ft long gas ignition zone and the dust beds were placed at 125, 227 and 320 ft distance from the face.

The concentrations reported in the first two columns can be converted into layer densities by multiplying them by the gallery height of 8 ft, or 2.4 m. For example, for the last test in the table (NGFA56), 227 g/m2 (i.e. 93 g/m3 X 2.4 m) cornstarch was spread over the floor, and explosion removed 119 g/m2, corresponding to an entrainment fraction of 53%.

Tamanini measured the pressure development in the primary chamber as well as at four different stations along the gallery. The data from test NGFA56 is shown in Fig. 3.7. An examination of this pressure data reveals the following:

– There are no significant differences among the primary pressure pulses experienced in the primary chamber and at upstream stations in the gallery
– Primary pressure pulse travels downstream at approximately sound speed, and
– The period of the Helmholtz oscillations are substantially smaller than the period of the primary pulse.

These observations suggest that the free stream gas velocity in the gallery might be estimated using the acoustic approximation for a simple wave defined by the measured pressure pulse. The red curve superimposed, in Fig 3.7, on the pressure trace recorded at 8.0 m in the tunnel represents our approximation to the time resolved pressure data in the gallery. For cornstarch, (3.2) predicts a threshold velocity of 5.3 m/s. Resulting entrainment predictions are compared to the

---

[4] Dynamic pressure impulse is the integral of dynamic pressure over time.

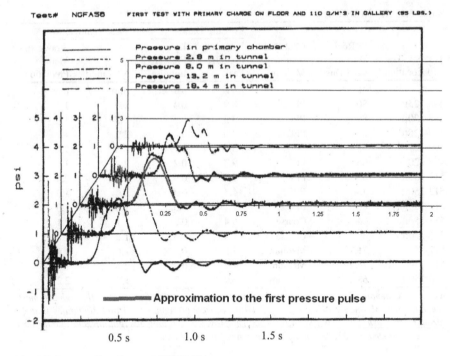

Fig. 3.7 Pressure data from test NGFA56

test data in Fig. 3.8. The abscissa in Fig. 3.8 corresponds to the peak pressure Tamanini recorded at the 8 m station. Considering the uncertainty due to gas velocity estimation, the agreement between the data and predictions is encouraging.

3.5   Estimation of the Dust Entrainment Caused by Primary Event Scenarios

Previous section showed that when the free stream velocity near the dust layer is known as a function of time, the new strawman method provides a reasonable representation of the dust removal data. When applying the strawman method for industrial hazard evaluation, temporal and spatial variation of the free stream velocity over the dust layer needs to be considered. This can be done with detailed computer modeling or using a simplified approach.

To demonstrate the latter approach, Project Panel task group selected the following typical examples:

Example A: Catastrophic burst of an indoor equipment (e.g. dust collector), and
Example B: Dust deflagration venting from a room into another room.

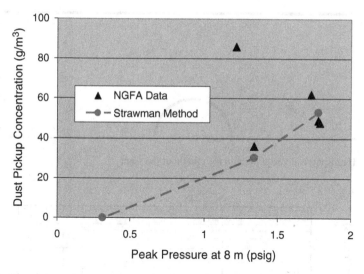

Fig. 3.8 Comparison of the cornstarch entrainment predictions with NGFA data

The dust entrainment in scenario A is driven by a blast wave, while that in Scenario B is driven by the vent discharge flow.

EXAMPLE A: Catastrophic burst of an indoor equipment

Consider a 74 ft3 dust collector sitting on the floor, in the middle of a plant bursts at 0.5 barg. The floor of the plant is covered with a thick layer of aluminum dust made up of 100 micron spherical particles. In this scenario, the Baker-Strehlow method[5] for bursting spheres can be used. If the burst is caused by an internal deflagration, the average temperature inside the enclosure depends on the availability of venting prior to burst. In this analysis, the enclosure contents were assumed to be at the ambient temperature and have the specific heat ratio of 1.4 prior to the burst, since this produces results that are more conservative. Effects of any combustion or shock wave reflections after the burst are ignored. For the burst pressure of 0.5 barg, Baker-Strehlow method predicts a maximum side-on pressure of 0.22 barg at the burst surface, approximately one-half of the burst pressure.

Since the dust collector is sitting on the floor, we can approximate the geometry as a hemisphere as shown in Fig. 3.9. The hemispherical treatment accounts for the confinement of the blast wave due to floor by doubling the dust collector volume, but ignores effects such as oblique and mach reflections. For the dust collector volume of 74 ft3, then the hemisphere radius is 1 m.

Figure 3.10 shows the peak pressure field predicted by the Baker-Strehlow method. The peak velocity field is ideally calculated using the shock relations.

---

[5] See, for example, CCPS (2010) for a calculation procedure.

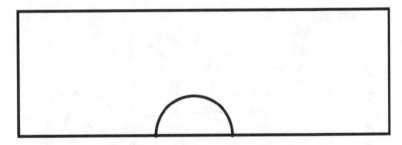

Fig. 3.9 Hemispherical vessel burst in the middle of the plant

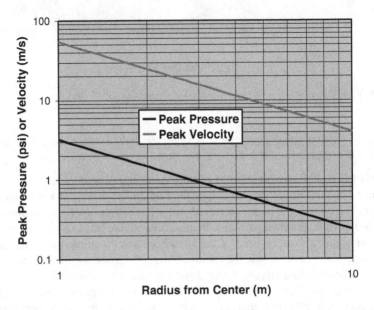

Fig. 3.10 Peak pressure and velocity fields created by a 74 ft3 hemispherical enclosure burst at 0.5 bar pressure

However, since the peak pressures are low, acoustic approximation is adequate in this case:

$$U = \frac{\Delta P}{\rho \cdot a} \tag{3.3}$$

Where:

U=U(r): peak free stream velocity (m/s) at radius r
$\Delta P = \Delta P(r)$: peak pressure rise (Pa) at radius r
ρ: density of air (1.2 kg/m3), and
a=340 m/s speed of sound in air.

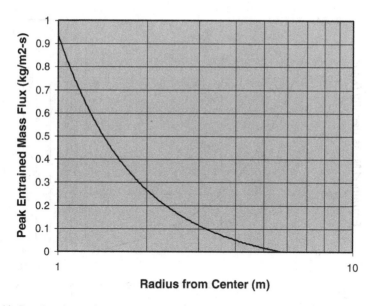

Fig. 3.11 Local peak entrainment mass flux for the 100-micron Aluminum dust

The peak velocity field calculated using the acoustic approximation is also shown in Fig 3.10. The peak velocity is seen to be equal to the 7.5 m/s threshold velocity, for the 100-micron Aluminum dust example above, at a radius of 5.6 meters. Therefore, beyond the 5.6 m "threshold radius[6]", our methodology predicts no entrainment. Peak velocity never exceed 54 m/s, a value determined strictly by the burst pressure.

Fig. 3.11 shows the local peak entrainment flux calculated using Equation (30).

At a given location, both the overpressure and velocity suddenly jump from zero to their respective peak values, and start decaying exponentially. After a finite period both the overpressure and velocity go through zero and change sign. For the sake of simplicity here, we will assume the peak overpressure and velocity at a given radius remain constant for a finite period. Its duration can be estimated from the peak overpressure and impulse values predicted by the Baker-Strehlow model. Assuming a triangular waveform:

$$\text{Duration, } \Delta t \ = 2^* \ \text{Impulse}/(\text{Peak Overpressure}) \tag{3.4}$$

It should be pointed out that this is a very conservative assumption for the dust entrainment calculations. For the present example, a duration $\Delta t = 0.0028$ sec is predicted.

---

[6] The threshold radius scales with the cube root of the bursting enclosure volume, and can conceivably be utilized to establish safe separation distances.

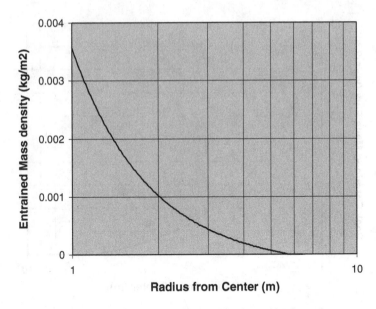

Fig. 3.12 Local entrained mass per unit area for the 100-micron Aluminum dust

Figure 3.12 shows the local entrained mass per unit area for the 100-micron Aluminum dust, and is calculated by multiplying the flux shown in Fig. 2.8 with the 0.0028 second duration.

Total entrained mass is calculated by integrating the values shown in Fig. 3.12 over the radius and is calculated to be only 19 grams. In other words, the new strawman methodology predicts no secondary explosion hazard for this scenario.

EXAMPLE B: Dust deflagration venting from a room into another room

Consider a 100 m3 primary enclosure communicating with the rest of the building through a 6′ by 8′ opening. The floor of the building is covered with a thick layer of aluminum dust made up of 100 micron spherical particles. In this example, we will estimate how much dust can become airborne if a 1 psi (6895 Pa) overpressure deflagration occurs in the primary enclosure.

For the sake of simplicity and conservative results, vent discharge is treated as a quasi-steady[7] turbulent round momentum jet, hugging the floor at the jet axis. A conceptual sketch is provided in Fig. 3.13. Since the vent opening is adjacent to the floor, the discharge jet takes the form of one-half of an axial jet discharging from an outlet twice as large. Air entrainment occurs almost exclusively along the surface of a half cone, and the maximum velocity remains close to the surface. Effects of surfaces other than the floor are ignored. Flame is assumed to exit the

---

[7] Temporal evolution process of the vent discharge jet was not considered to keep the analysis simple for the end user.

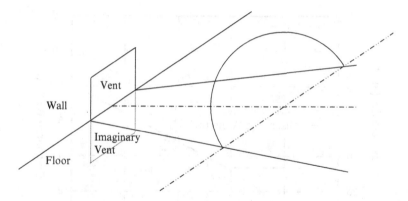

Fig. 3.13 Conceptual schematics of vent discharge treated as half a round floor jet

primary explosion enclosure after all the contents have been consumed. This is a conservative assumption consistent with rear ignition.

The peak discharge velocity can be calculated from the Bernoulli's equation. For the most conservative discharge coefficient of unity, exit velocity is

$$Uo = \sqrt{\frac{2 \cdot \Delta P}{\rho}} = \sqrt{\frac{2 \cdot 6895}{1.2}} = 107 \cdot m/s \tag{3.5}$$

The opening has a cross-sectional area of 48 ft2 or 4.5 m2. To simulate the effect of the floor, a mirror image imaginary vent below the floor is considered and the discharge area is doubled. Therefore, equivalent discharge diameter, Do is 3.37 meters.

Variation of the peak axial velocity in a free jet is approximately represented with four zones (see, for example, ASHRAE 2005). Zone 1 is a short core, extending several exit diameters, in which the maximum velocity of the flow remains practically unchanged. Zone 3 represents fully established turbulent flow that may be 25 to 100 diameters long. Transitioning between Zones 1 and 3, there is a short Zone 2. Zone 4 is a zone of jet degradation where the flow velocity becomes comparable to that for preexisting turbulent fluctuations and convective currents.

In this example, peak axial velocity decay with axial distance X is represented by the equation:

$$U = Uo \min[1, 6.2D/X] \tag{3.6}$$

It should be noted that (3.6) represents Zones 1 and 3, while ignoring Zones 2 and 4.

If we assume a top-hat velocity profile, the width, W, of the jet hugging the floor at distance X has to be

$$W = Do \, Uo / U \tag{3.7}$$

to preserve the momentum.

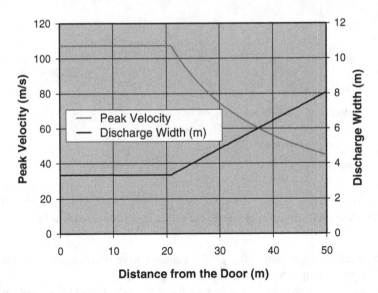

Fig. 3.14 Peak velocity field, and the discharge width created by the Example worked out for Scenario B

Assuming all vent discharge occurs at the 107 m/s exit velocity, the discharge duration is limited with the mixture in the primary enclosure, and can be calculated as:

$$\Delta t = \frac{7}{8} \cdot \frac{100m^3}{4.5m^2 \cdot 107m/s} = 0.18 \text{ seconds} \tag{3.8}$$

if there are no other vents available for discharge. The factor 7/8 accounts for the residual burnt mixture inside the primary enclosure. The discharge duration is controlled by the volume of the primary enclosure. Alternatively, Equation 6.3.5.5. in NFPA 68 predicts a discharge duration of 1.0 seconds for Pmax = 8 barg.

Figure 3.14 shows the peak velocity as a function of distance from the vent opening. The magnitude of the maximum velocity is 107 m/s and is determined by the pressure the primary deflagration enclosure can withstand. The velocity decay with distance is seen to be slower in this Scenario, owing primarily to the direction-ality of the discharge. The velocity decay is controlled by the area of the vent opening. However, unlike the burst scenario which sends a blast wave in all directions, the vent discharge affects dust layers only over a narrow path of width W along the jet axis.

Figure 3.15 is a plot of the local peak entrainment flux calculated using (3.2).

Figure 3.16 shows the local entrained mass per unit area for the example worked out in this section, and is calculated by multiplying the flux shown in Fig. 2.12 with the 0.18 second duration. It is instructive to note that the simplified analysis presented here predict, for the worked out example, entrainment of 0.5 kg/m2 at

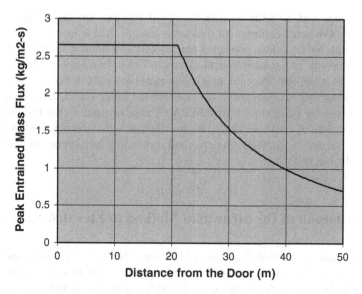

Fig. 3.15 Local peak entrainment mass flux for the Example worked out for Scenario B

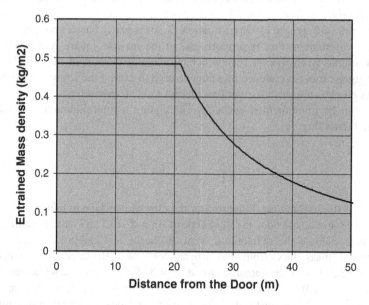

Fig. 3.16 Local entrained mass per unit area for the Example worked out for Scenario B

the vent opening, and 0.13 kg/m2 at 50 m distance from it. Recognizing the fact that layer thickness allowance in current NFPA standards work out to be approximately 1 kg/m2, method predictions correspond to entrainment fraction of to 50% at the vent opening, and 13% at a distance of 20 meters, for the threshold layer thicknesses. Predicted entrainment fraction is less for thicker layers.

The simplified equations presented here predict entrainment for distances up to 300 m, and result in a total entrained mass of 173 kg. In reality, the length of the discharge jet is limited either because it impinges on a wall, or because it detaches from the floor due to buoyancy. To estimate the latter effect, flame projection distance given by Equation 8.8.2 of NFPA 68 may be used. For a 100 m3 room, single vent, and a K coefficient of 10 for metal dusts, a flame projection distance of 46.4 m is calculated. The Strawman method results in a total entrained mass of 68 kg up to 46.4 m distance.

3.6 Extension of the Strawman Method to Elevated Surfaces

The data used to develop Equation (3.30) were obtained in experiments with airflow over dust deposits covering relatively large surface areas. Equation (3.30) should also be applicable to dust deposits with streamwise dimension exceeding 2 m long. However, most elevated surfaces such as typical I-beams and roof support structures have characteristic dimensions that are substantially smaller than 2 meters.

The work of Batt et al (1995), Equation (3.30) is partly based upon, determined that dust entrainment flux is proportional to the inverse square root of the dust deposit length. Even though such a behavior is not evident in the NIOSH data, inverse square root dependence rule may be used to extend the strawman method to deposits on elevated surfaces, until validation tests are performed.

This can be implemented easily by calculating a multiplication factor α for entrained mass flux:

$$\alpha = \sqrt{\frac{2}{L}} \qquad \text{for L<2m} \qquad (3.9)$$

Where L is the streamwise dimension of the elevated surface in meters.

As an example, consider the dust deposit on a 4″ (0.1 m) wide I-beam placed in the near field of the deflagration vent discharge in Example B, above. Then entrainment mass flux from the I-beam can be calculated by multiplying Equation (3.30) with the factor factor $\alpha = 4.5$. Since earlier calculations based on Equation (3.30) showed that dust entrainment from floor deposits is approximately 0.5 kg/m2, the dust entrainment from the I-beam will be approximately 2.25 kg/m2. Primary deflagration will remove top 2.25 mm of the dust deposit on the I-beam, if the bulk density of the dust deposit is 1000 kg/m3. This corresponds to an entrainment fraction of 71% if the dust layer thickness is 1/8″, or 100% if the dust layer thickness is smaller than or equal to 2.25 mm.

Chapter 4
Validation Plan

As discussed in the previous chapter, the new strawman method is essentially comprised of two major components:

a. selection of the types, magnitudes and durations of the maximum credible disturbances
b. calculation of the mass of the dust entrained from the deposits.

The first component depends on the primary event scenarios that are credible for the specific occupancy. The two most commonly encountered scenarios have been developed in the previous chapter. It is anticipated that additional scenarios will be identified and developed as the new strawman method is being adopted into the combustible dust occupancy standards. When developing the two scenarios presented in this report, it was possible to identify and utilize well-known simple tools. The same may be true for the new scenarios as well.

The heart of the strawman method is the entrainment mass flux correlation. Even though it was predominantly based on high quality data, additional tests will be useful to build confidence on its applicability to a wide range of combustible dusts covered in NFPA 61, NFPA 484, NFPA 654, and NFPA 664. Depending on how well the correlation compares to the new data, its coefficient and exponents may be adjusted as necessary.

4.1 Uncertainties in the Mass Flux Correlation

The high velocity portion of the mass flux correlation was originally obtained by Batt et al (1999) for sand particles. The two types of sand used in these tests had mass median particle size of 180 microns, and 250 microns. Sand particles are compact in shape and have a particle density of 2600 kg/m3. The free stream velocity used in these tests were nearly constant throughout each test and ranged from 30 to 100 m/s.

E.A. Ural, *Towards Estimating Entrainment Fraction for Dust Layers*,
SpringerBriefs in Fire, DOI 10.1007/978-1-4614-3372-9_4,
© Fire Protection Research Foundation 2011

This correlation also provided favorable prediction of the NIOSH data for coal dust / rock dust mixtures with coarse as well as fine particle size (i.e. 80% smaller than 74 microns). NIOSH uses particle density of 1330 kg/m3 for coal dust and 2750 kg/m3 for rock dust. The free stream velocities recorded in NIOSH tests were highly transient pulses induced by the primary explosion and ranged from 0 to above 200 m/s.

As also discussed in the previous chapter, a limited set of cornstarch entrainment data were available from the NGFA tests, and provided a good agreement with the mass flux correlation. It should be noted that cornstarch is made up of much finer particles that have a mass mean size of 15 microns and particle density of 1550 kg/m3. The free stream velocities recorded in NGFA tests were transient pulses induced by the primary explosion and were estimated to range from 0 to 27 m/s.

Additional tests employing lower density (e.g. wood dust with approximately 500 kg/m3 particle density), and higher density (e.g. iron dust with approximately 7800 kg/m3 particle density) will be beneficial.

The mass flux correlation employed in the strawman method accounts for the effect of nonspherical shaped particles, through a correction factor for the threshold velocity. This effect was introduced in an ad hoc fashion, and needs to be validated. The form of the mass flux equation reveals that nonspherical shape effect becomes more pronounced in lower free stream velocities. This behavior intuitively makes sense but needs to be validated with additional tests in the next phase of this project. Furthermore, the proposed form of the mass flux equation implies that entrainment for nonspherical shaped particles will initiate at lower free stream velocities and proceed at a higher rate than that for spherical particles. This behavior also needs to be tested and either validated or adjusted appropriately.

Finally, the extension of the mass flux correlation to elevated surfaces with small streamwise dimension needs to be validated.

4.2   Recommended Test Arrangement

For the sake of simplicity and data reliability, a one-dimensional large-scale test set up is recommended. Similar to the NIOSH tests, a primary deflagration or a blast wave can be set up at the closed end of a gallery. NIOSH tests were performed in a full-scale mine with a rather large cross-sectional area (20 ft by 7 ft). NGFA tests, on the other hand employed a test set up that was 8 ft by 8 ft in cross-section. It is conceivable that testing in even a smaller scale can produce good results. However, the minimum acceptable dimensions will depend how the air pulse is generated and its duration. A driver system producing a weaker negative phase will be desirable. It is also recommended that the construction of the walls and floor of the selected facility should not be conducive to transmission of mechanical shocks that can be caused by the initiating event. The length and the boundary conditions of the test facility are important factors that must be considered when selecting or constructing a test facility.

Carefully formed dust beds can be placed on the floor and short span elevated surfaces at sufficiently far downstream locations. The free stream velocity and static pressure transient need to be recorded at each dust bed station. Free stream velocity should be measured near the center of the cross-section, using calibrated probes that have response time fast enough to capture the time variation of the free stream velocity faithfully. If a primary deflagration is used to create the velocity transient, care must be exercised to ensure that flames do not ignite the dust beds.

In these tests, the peak free stream velocity should be varied from slightly above the threshold value to at least 100 m/s. At least three replicate tests for each condition will be desirable.

As in NIOSH tests, total dust entrainment can be measured from the difference in the dust bed thicknesses before and after the primary event. Preferably, the dust deposits are weighed before and after each test, resulting in a determination of the mass removal. If the selected facility has the capability to map the time resolved suspended dust concentration and velocity over the flow field, dust entrainment mass flux can be deduced.

It may be possible to get these tests performed at NIOSH for a little or no cost to the Fire Protection Research Foundation under the NORA Intramural Research Program.

4.3   Recommended Test Matrix

The following test matrix is envisioned.

- Peak free stream velocity: 15 m/s, 30 m/s, 60 m/s, and 100 m/s or higher.
- Velocity pulse duration: preferably long, and to be decided based on available test setup details.
- Combustible dusts: wood dust, iron dust, tissue dust (or flocking fibers), aluminum flakes.
- Dust layer thickness: two to three thicknesses to be selected based on velocity pulse duration and particle density.
- Elevated surface span: representative dimensions of I-beam and roof support structures. 4″, 12″ and 24″ are recommended.

By prudent placement of the beds, floor dust entrainment and elevated surface entrainment data can be generated simultaneously in each test. The test matrix may be modified as data becomes available.

Appendix A
Ad Hoc Methods to Characterize Material Dustiness and Entrainability

A menu of simple tests which have been devised to classify or screen powder properties are introduced. These tests, intuitively believed to relate to the problem at hand, can be grouped into the following four categories: (1) particle characterization; (2) cohesion tests; (3) terminal velocity tests, and (4) dispersibility tests.

A.1 Particle Characterization

Particle Size and Shape:

In most powders, there is a wide gamut of particle sizes and shapes. For irregular particles (as would be the case for most ground powders), the shape and size are difficult to characterize. Therefore, the approach has been to define a size based on a parameter of great importance for a given application. For example, for paint pigments, the projected area; for chemical reaction the total surface area is considered important; thus, particles are characterized by the projected area diameter and the surface diameter, respectively. For the dust dispersion problem, an aerodynamic drag diameter should be appropriate to characterize the cloud behavior. For the dust pick-up, however, the choice of characteristic size is not as clear.

Particle Density:

The density of individual particles is determined by using pyknometers based on gas or liquid displacement by known weight of the solid particles. If the particles are porous, the liquid will not normally enter the pores because of the surface tension. Therefore, density values measured by liquid and gas pyknometers differ and are called envelope density and true density, respectively. For the dust entrainment problem, envelope density is probably more appropriate, making liquid pyknometer methods preferable.

E.A. Ural, *Towards Estimating Entrainment Fraction for Dust Layers*,
SpringerBriefs in Fire, DOI 10.1007/978-1-4614-3372-9,
© Fire Protection Research Foundation 2011

Bulk Density:

The bulk density of a material is the mass per unit volume of many particles of the material. For most powders, bulk density depends on the degree of packing, so that it is customary to define an "aerated (loose) bulk density," and a "packed bulk density." The aerated bulk density is generally determined by gently filling a container, whereas the packed bulk density applies to powder that has been packed by means of vibrating the container typically for 5 minutes. In bulk solids handling, bulk density is believed to have no effect on material flowability. In entrainment phenomenon, however, bulk density may play an important role.

Carr Compressibility:

Compressibility is defined as the ratio of the difference between packed and aerated densities normalized by the packed density (ASTM D6393). Generally, the more compressible the material, the less flowable it will be. According to Carr (1965), the borderline for free flowing to non-free flowing is about 20% compressibility. In terms of the microscopic properties, compressibility depends on particle shape, size distribution, deformability, surface area, cohesion, and moisture content.

Carr Uniformity Coefficient:

This is a measure of the particle size distribution. Carr (1965a) and ASTM D6393 define it as the ratio of the width of the sieve opening that will pass 60% of the sample by the width of the sieve opening that will pass 10% of the sample. The conventional wisdom in bulk solid handling is that the more uniform a mass of particles is in both size and shape, the more flowable it is likely to be.

Moisture Content:

This is recognized as one of the important parameters affecting the entrainment of dust, since both electrostatics and cohesive forces are strongly dependent on the moisture content. ASTM has written extensive test procedures to determine the moisture content of a number of substances, most of which are based on the weight loss after drying the sample at a specified temperature for a specified period.

A.2   Cohesion Tests

Tackiness Test:

A rough test for tackiness is to try to form a ball of the material by rolling it between the palms of the hands. If the ball falls apart, the material is not tacky (Kraus, 1980). Another qualitative test for tackiness is to check whether the material, smeared onto a metal surface, will build up so that it cannot slide off.

Carr Cohesion Test (ASTM D6393):

A somewhat more quantitative test has been used by Carr (1965 a). This procedure for apparent surface cohesion involves determining the retention of material on a nest of 60 (220 um), 100 (147 um) and 200 (74 um) mesh screens, plus a bottom pan. A 2 gram sample presifted to minus 200 mesh is placed on the top screen (60—mesh) and the system is vibrated for 20 to 120 seconds depending on the bulk density of the material. At the end of the vibration period, material left on each screen is weighed, and based on the weight distribution with respect to the screen, a cohesion index is calculated.

Carr Angle of Repose (ASTM D6393):

This is the angle between the horizontal and the natural slope formed by a pile of material dropped from some elevation. This angle, to a degree, should depend on the details of how the material is being dropped. Although angle of repose is frequently used to characterize materials, no standard test procedures exist. The angle of repose is a measure of friction between particles and is related to the potential flowability of material. At the microscopic level it is expected to depend on particle shape, size, porosity, cohesion, fluidity and surface area.

Carr Angle of Fall (Angle of Slide):

After carefully building up the heap for the angle of repose measurement on a metallic plate, a standard weight is dropped onto the plate from a prescribed height a predetermined number of times. A new angle of repose results from the jarring. This angle is called the "angle of fall." The flattening of the heap may result from flow of the particles down the slope or the collapse of the pile. The latter may indicate entrapped air within the heap. The angle of fall depends on the fluidity, shape, size and uniformity of particles, entrapped air in the heap and cohesion.

Carr Angle of Difference (ASTM D6393):

This is the difference between angle of repose and the angle of fall, and it is believed to relate to the potential for flooding or fluidity of a material. The angle of difference depends on the fluidity, surface area and cohesion.

Carr Angle of Spatula (ASTM D6393):

This angle is measured by sticking a spatula with a sufficiently wide blade into the bottom of a mass of the dry material and then lifting it straight up. A free—flowing material will form a well—defined angle over the spatula, whereas a non—free-flowing material will form a number of irregular angles over the blade. Carr (1965a) recommends the following procedure for determining the angle of spatula: "First, angles to the horizontal are measured and an average value taken. Then the spatula is tapped gently, producing a lower angle or angles of spatula that are measured and again averaged. The average of the two averages is termed the angle of spatula."

Angle of spatula is determined by cohesion, particle surface area, size, shape, uniformity, fluidity, porosity and deformability.

Shear Cell Test (ASTM D6128):

The shear cell (Jenike, 1964) consists of two rings, one sitting on top of the other, one secured on a table. The volume inside the two rings is filled with powder to be tested, which is compressed axially by loading the cover plate. The test determines the horizontal force required to move the upper ring as a function of the compressive load. Based upon the cross- sectional area, a shear stress is calculated, believed to represent the magnitude of the cohesive and frictional forces between dust particles.

A.3 Terminal Velocity Tests

The terminal velocity of a particle or terminal velocity distribution of a powder is a well-defined and measurable parameter. The reason for the inclusion of these tests in the intuitive approach category is the intuitive relationship between the dust dispersibility and the terminal velocity. The measurement techniques of terminal velocity of solid particles suspended in liquids are well developed and are frequently used to determine particle size distributions. A good review of this subject is given by Allen (1975). However, there are only a few methods for terminal velocity measurements in air, mainly due to the difficulty of dispersing powders.

Sedimentation Technique:

In this method, the powder is initially dispersed as a thin cloud near the top of a vertical tube containing otherwise quiescent air. The dust cloud is allowed to settle in the tube, and the amount of dust collected at the bottom is measured as a function of time. The height of the tube and the time measured to collect various mass fractions are used to calculate the terminal velocity distribution. The classical example of this method is the micromerograph originally developed by Eadie and Payne (1956) to measure particle size distribution in the sub-sieve range. The settling velocity apparatus developed by Ural (1989) is a modified and larger scale version of the micromerograph.

Elutriation Technique:

This technique, mostly used for non-cohesive granular material, is based on introducing particles into a vertical tube containing an upward air flow adjusted to the point where the particles are held in a nearly stationary position (NMAB, 1982).

Centrifugal Technique:

The classical example of this technique is the Bahco microparticle classifier, described, for example, by Allen (1975). This apparatus has been widely used since its adoption by the ASME Power Test Codes (PTC28) as a standard method of measuring terminal velocity for the design and evaluation of dust collection equipment. The sample is introduced into a spiral shaped air current created by a hollow disk rotating at 3500 rpm. Air and dust are drawn through the cavity in a radially inward direction against centrifugal forces. Separation into different size fractions is made by altering the air velocity. The apparatus is calibrated using a standard dust with a known terminal velocity distribution, provided by ASME.

A.4   Dispersibility Tests

Dispersibility can vaguely be defined as the ease of a powder to mix with air. If the ratio of the mean particle spacing to particle size is much larger than one, the mixture is referred to as a dust cloud, whereas if the ratio is of the order of one, the term aerated (or fluidized) powder is used.

Simple qualitative tests are performed to determine aeration and de—aeration characteristics of powders. A few examples given in Kraus (1980) are as follows:

— Shake the material in a partially filled container and note if it swells in volume due to aeration
— Stand the container of aerated material on a flat surface and note the time it takes to de-aerate of its own accord
— Rapidly tap a container of aerated material at a constant rate and note the time it takes to de-aerate
— Shake the material in a partially filled cardboard tube having a taped hole in the side near the bottom. Stand the tube vertically on the edge of a table, remove the tape and note the trajectory of the stream of material. A fluidized material will issue like water; a dead material may just ooze and plug the orifice.

A simple and somewhat quantitative procedure for characterization of dispersibility is described by Carr (1965a) and ASTM D6393. The apparatus used consists of a 4—in. I.D. plastic cylinder 13 in. long, supported vertically from a ring—stand 14 in. above a 14 in. diameter watch-glass. A 10 gram sample of material is dropped "en-masse" through the cylinder from a height of 24 in. above the watch glass. Material remaining on the watch glass is weighed, and the difference from the initial mass yields the amount dispersed during the fall.

A classical experiment of dispersibility was performed by Andreasen (1939). Two cubic centimeters of powder were poured through a narrow slit into a vertical

Table A-1 Dispersibility of some powders measured by Andreasen (1939)

	Particle Radius Limits um	Dispersibility (%)
Lycopodium	12	100
Wood charcoal dust	0–25	85
Wood charcoal dust.	0–7	23
Aluminum powder	0–15	66
Talc	0–20	57
Carbon black	0–.15(?)	147
Potato starch	0–35	27
Graphite dust	0–25	17
Pulverized slate	0–25	13
Cement	0–45	5.5
Prepared chalk	0–6	1.5
Polydisperse silica dust (coarse)	–	21
Polydisperse silica dust (fine)	–	8
	11.5	68
Isodisperse silica dust	8	83
	5.6	145
	7	50
Porcelain dust with fine fractions	2.7	52
	1.1	21
	0.145	12
Porcelain dust without fine fractions removed	–	5

tube of 250 cm height and 14.5 cm diameter. The particles were separated to some extent as they fell through the air, and the percentage of powder, which had not settled to the bottom of the tube in 6 seconds, was determined. Since the individual particles could not have reached the bottom in this time, the author assumed that this figure represented the percentage of dispersed powder, which he called dispersibility. Andreasen's experiments have been criticized on the grounds that the unsettled fraction also contained small aggregates. Some of his data is given here as Table A-1.

ASTM developed two standards for characterization of the dustiness of powders. ASTM D547-41(1980) Test Method for Index of Dustiness of Coal and Coke (Withdrawn 1986) was used to evaluate the dustiness for coal and coke samples. ASTM D4331-84, Practice for Measuring Effectiveness of Dedusting Agents for Powdered Chemicals (Withdrawn 1988), was used to assess the effectiveness of dedusting agents for powdered chemicals.

The apparatus for ASTM D547 consisted of a metal cabinet 5 ft in height and 18 in. square, inside dimensions, arranged with a cover and three horizontal slides and a drawer at the bottom. The top slide was inserted 12 in. below the top of the cabinet and formed the compartment into which the sample (50 lb) was placed. The other two slides were inserted together 24 in. above the bottom (24 in. below the top slide) and were used to collect the settled dust after 2 and 10 minute intervals. Immediately before the test, the upper slide was in the inserted position, whereas the lower slides were in withdrawn position.

The test started with the quick withdrawal of the upper slide, allowing dust to drop into the bottom drawer. Exactly 5 seconds later both lower slides were inserted. After the slides have stood undisturbed for 2 minutes from the time of insertion, the upper one of the two lower slides was withdrawn. The second lower slide was withdrawn after an additional 8 minutes has elapsed. Two dustiness indices (coarse dust and float dust) were defined, based on the weight of the samples on these two slides. ASTM requires a better than 20% reproducibility for the measurements.

ASTM Test D4331 used a 3 in. plastic tubing (2 7/8 in. I.D.) 18 in. long as a fluidizing environment. The tube was held vertically while a constant air flow rate corresponding to 6 cm/s bulk velocity in the tube was supplied from the bottom. The top was covered with a partially open (27%) cap. Also located at the top is a 5/16 in. I.D. sampling probe with a constant suction velocity pump through a dust filter. During the test, the tube was vibrated at 29,000 rpm. The standard called for a 200 g sample and 20 minute exposure. The amount of dust collected in the filter during this time was used as a relative measure of dustiness.

Another widely used method of dispersibility characterization is to dump a known quantity of material into a volume equipped with a high volume air sampler at the top. The ratio of the mass of the powder collected by the sampler to the mass of the material handled is used as a measure of the dustiness of the material. This method is used for comparison purposes, since the results are dependent on the test apparatus and procedures. An example of this method can be found in Lundgren and Rangaraj (1986).

Recently VDI 2263 (part 9) published a standard test method to determine dustiness of bulk materials. A metering device conveys the sample into a measuring chamber at constant volume flow rate. The concentration of the dust cloud forming in the measuring chamber is recorded as a function of time. The standard calls the average of the measured dust concentration (averaged over the test duration) the dustiness coefficient, S. Material is classified into one of the six dustiness groups based on the value of the dustiness coefficient as shown below.

VDI Dustiness Group	Dustiness Index (g/m3)
1	LT 1
2	1 to 5
3	5 to 10
4	10 to 20
5	20 to 50
6	GT 50

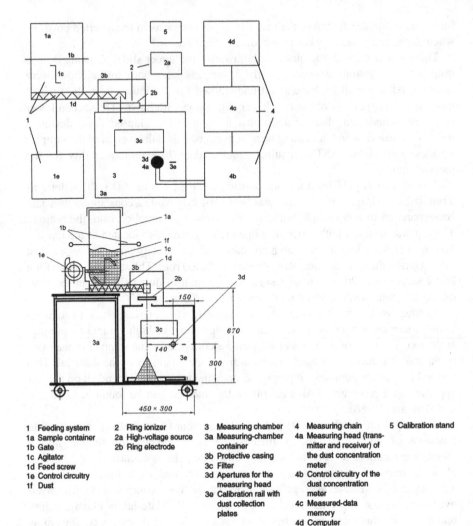

1	Feeding system	2	Ring ionizer	3	Measuring chamber	4	Measuring chain	5	Calibration stand
1a	Sample container	2a	High-voltage source	3a	Measuring-chamber container	4a	Measuring head (transmitter and receiver) of the dust concentration meter		
1b	Gate	2b	Ring electrode	3b	Protective casing				
1c	Agitator			3c	Filter	4b	Control circuitry of the dust concentration meter		
1d	Feed screw			3d	Apertures for the measuring head				
1e	Control circuitry			3e	Calibration rail with dust collection plates	4c	Measured-data memory		
1f	Dust					4d	Computer		

Fig. 1 Block diagram and mechanical structure of the measure system

Appendix B
Alternative Mechanism on Dust Cloud Generation

A Panel member requested consideration of vibration or "Elastic Rebound" mechanism, shown below as promoted in some NFPA training publications. Even though this topic is beyond the scope of this project, an additional literature survey is conducted upon the Panel request.

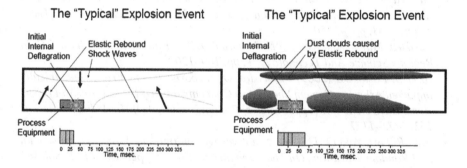

Search was performed by querying Science Direct for the following keywords:

dust cloud generation mechanical vibration
"dust layer" "mechanical shock"
"dust cloud" "mechanical shock"
"dust cloud" "mechanical vibration"
"dust layer" "mechanical vibration"
"powder" "mechanical vibration"
"powder layer" "vibratory"
"dust layer" "vibratory"
"powder layer" "mechanical impact"
"powder layer" "mechanical shock"
"dust cloud" "mechanical impact"
"powder cloud" "mechanical impact"
"powder cloud" "mechanical shock"

Only Relevant mention found in a 1984 Eckhoff article:

HOW CAN DUST CLOUDS BE GENERATED AND IGNITED IN LARGE SILO CELLS? This question certainly has more than one answer. In case the main material to be stored in the silo is in itself sufficiently fine to give explosible clouds in air, explosible dust clouds are most likely to arise, at least transiently, somewhere in the silo whenever new material is discharged into it, whether pneumatically or mechanically. If the main material is coarse, such as grain, explosible dust clouds may be generated by unburnt dust being blown into the silo by preceding explosions elsewhere in the plant. Dust could, for example, be injected through the various openings close to the silo top. Injection through the hopper exit at the bottom seems a more unlikely scenario. Another process of dust cloud generation could be that dust layers, which have accumulated on the inside of the silo wall and roof, are disturbed and dispersed into a cloud by air blasts or mechanical vibrations induced, for example, by preceding explosions elsewhere in the plant.

Other interesting but hardly useful papers included:

- *P. C. Arnold And K. S. Kaaden (1977), "Reducing Hopper Wall Friction By Mechanical Vibration," Powder Technology, 16, pp. 63–66*
- *Diego Barletta, Giorgio Donsi, Giovanna Ferrari, Massimo Poletto, Paola Russo (2008), "The Effect Of Mechanical Vibration On Gas Fluidization Of A Fine Aeratable Powder," Chemical Engineering Research And Design, 86 pp. 359–369.*
- *Edward K. Levy, Brian Celeste (2006), "Combined Effects Of Mechanical And Acoustic Vibrations On Fluidization Of Cohesive Powders," Powder Technology, 163, pp. 41–50.*
- *Norikazu Maeno And Kouichi Nishimura (1979), "Fluidization Of Snow," Cold Regions Science And Technology, 1, pp. 109–120.*
- *Jozef S. Pastuszka (2009) "Emission Of Airborne Fibers From Mechanically Impacted Asbestos-Cement Sheets And Concentration Of Fibrous Aerosol In The Home Environment In Upper Silesia, Poland," Journal Of Hazardous Materials 162, 1171–1177.*
- *A.W. Roberts And O. J. Scott, "An Investigation Into The Effects Of Sinusoidal And Random Vibrations On The Strength And Flow Properties Of Bulk Solids," Powder Technology, 21. pp. 45–53.*
- *P. H. Gregory And Maureen E. Lacey (1963), "Liberation Of Spores From Mouldy Hay," Trans. Brit. Mycol. Soc. 46 (1),73-80.*
- *Gyorgy Ratkai (1976) "Particle Flow And Mixing In Vertically Vibrated Beds," Powder Technology. 15, pp. 187–192*
- *Th. Kollmann And J. Tomas (2001) "Vibrational Flow Of Cohesive Powders," Handbook Of Conveying And Handling Of Particulate Solids, pp. 45–56.*
- *Chunbao Xu, And Jesse Zhu (2005), "Experimental And Theoretical Study On The Agglomeration Arising From Fluidization Of Cohesive Particles—Effects Of Mechanical Vibration," Chemical Engineering Science 60 pp. 6529–6541*

Interestingly, the last paper listed above concludes:

The experimental results prove that mechanical vibration can significantly reduce both the average size and the degree of the size-segregation of the agglomerates throughout the whole bed. However, the experiments also reveal that the mean agglomerate size decreases initially with the vibration intensity, but increases gradually as the vibration intensity exceeds a critical value. This suggests that the vibration cannot only facilitate breaking the agglomerates due to the increased agglomerate collision energy but can also favour the growth of the agglomerates due to the enhanced contacting probability between particles and/or agglomerates.

The so-called "Elastic Rebound" mechanism was also analyzed by mating a bursting sphere model with a single degree of freedom structural dynamic response model. For the typical "Spec building" structural parameters, it was demonstrated that the subject mechanism does not appear to be credible. This analysis will be published elsewhere.

Bibliography

Akiyama, T. and Y. Miyamoto (1989), "Solid Particle Transport from a Trough in a Wind Tunnel," Powder Technology, 57, pp. 235–240.

Allen, J.R.L. (1970), Physical Processes of Sedimentation, American Elsevier, Chapter 3.

Allen, T. (1975), Particle Size Measurement, Chapman and Hall, London.

Andreasen, A. (1939), Op. Cit. Fuchs (1964), Ref. 559.

Anonymous (1989), "Barriers Against Granary Blasts - White Oil and Soybean Oil Make Gains in What was Once the Province of Dust Collectors," Compressed Air Magazine, February issue, pp. 6–10.

Ansart, Renaud, Alain de Ryck, John A. Dodds (2009), "Dust Emission In Powder Handling: Free Falling Particle Plume Characterization," Chemical Engineering Journal, 152, pp. 415–420.

Ansart, Renaud, Alain de Ryck, John A. Dodds, Matthieu Roudet, David Fabre, François Charru (2009), "Dust Emission By Powder Handling: Comparison Between Numerical Analysis And Experimental Results," Powder Technology, 190, pp. 274–281

ASHRAE (2005), "Space Air Diffusion," 2005 ASHRAE Handbook Fundamentals, American Society of Heating, Refrigerating and Air-Conditioning Engineers, Chapter 33.

Bagnold, R A. (1941), The Physics of Blown Sands and Desert Dunes, Methuen and Co., London.

Batchelor, G.K, and Wen, C.S. (1982), "Sedimentation in a Dilute Polydisperse System of Interacting Spheres, Part 2, Numerical Results," J. Fluid Mech., v. 124, pp. 495–528.

Batchelor, G.K. (1972), "Sedimentation in a Dilute Dispersion of Spheres," J. Fluid Mech., v. 52, Part 2, pp. 245–268.

Batt, R. G. and A. A. Peabody (1995), "Dust Sweep-up Experiments," Defense Nuclear Agency Technical Report DNA-TR-94–117, November.

Boiko, V. M., Papyrin, A. N., Wolinski, H. and Wolanski, P. (1984), "Dynamics of Dispersion and Ignition of Dust Layers by a Shock Wave," Dynamics of Shock Waves, Explosions, and Detonations, (Bowen, J.R., Hanson, N., Oppenheim, A.K. and Soloukhin, R.I., Eds.), Progress in Astronautics and Aeronautics, AIAA, New York, v. 94, pp. 293–301.

Braaten, D. A., K. T. Paw and R. H. Shaw, "Particle Resuspension In A Turbulent Boundary Layer - Observed And Modeled," J. Aerosol Sci, Vol 21 No 5, pp 613–628, 1990

Bracht, K. and W. Merzkirch (1979), "Dust Entrainment In A Shock-Induced, Turbulent Air Flow," Int. J. Multiphase Flow, Vol. 5, pp. 301–312.

Cabrejos, Francisco J. and George E. Klinzing (1992), "Incipient motion of solid particles in horizontal pneumatic conveying," Powder Technology, 72, pp. 51–61.

Cabrejos, Francisco J., George E. Klinzing (1994), "Pickup and Saltation Mechanisms Of Solid Particles In Horizontal Pneumatic Transport," Powder Technology, 79, pp. 173–186.

Cabrejos, Francisco J., George E. Klinzing (1995), "Characterization of Dilute Gas-Solids Flows Using The Rescaled Range Analysis," Powder Technology, 84, pp. 139–156.

Calvert, G., M. Ghadiri, R. Tweedie (2009), "Aerodynamic Dispersion Of Cohesive Powders: A Review Of Understanding And Technology," Advanced Powder Technology, 20 pp. 4–16.

Carmichael, G.R. (1984), "The Effect of Shape on Particle Solids Flow," Particle Characterization Technology, (J.K. Beddow, Ed.), CRC Press, v. II, pp. 205–221.

Carr, R.L., Jr. (1965a), "Evaluating Flow Properties of Solids," Chemical Engineering, Jan. 18, pp. 163–168.

Carr, R.L., Jr. (1965b), "Classifying Flow Properties of Solids," Chemical Engineering, Feb. 1, pp. 69–72.

Cashdollar K L, Sapko M J, Weiss E S, Harris M L, Man C, Harteis S P, and Green G M (2010). Recommendations for a new rock dusting standard to prevent coal dust explosions in intake airways, NIOSH, RI 9679, 49 pp.

CCPS (2010) Guidelines for Vapor Cloud Explosion, Pressure Vessel Burst, BLEVE and Flash Fire Hazards, Wiley, section 7.4.

Chang, F.L., Sichel, M. and Kauffman, C.W. (1987), "Numerical Simulation of Flame Propagation due to Layered Combustible Dust," Chemical and Physical Processes in Combustion, 20th Fall Technical Meeting, Eastern Section: The Combustion Institute, No. 33.

Chein, R. and Chung, J.N. (1988), "Simulation of Particle Dispersion in a Two-Dimensional Mixing Layer," AIChE Journal, v. 34, No. 6, pp. 946–954.

Chi, D.N.H. and Perlee, H.E. (1974), "Mathematical Study of a Propagating Flame and Its Induced Aerodynamics in a Coal Mine Passageway," U.S. Bureau of Mines, RI 7908.

Chow, Judith C., John G. Watson, James E. Houck, Lyle C. Pritchett, C. Fred Rogers, Clifton A. Frazier, Richard T. Egami and Bridget M. Ball (1994), "A Laboratory Resuspension Chamber To Measure Fugitive Dust Size Distributions And Chemical Compositions" Atmospheric Environment Vol 28, No 21, pp. 3463–3481.

Cleaver, J.W. and Yates, B. (1973), "Mechanism of Detachment of Colloidal Particles from a Flat Substrate in a Turbulent Flow," Journal of Colloid and Interface Science, v. 44, No. 3, pp. 464–474.

Clift, R. (1978), Bubbles, Drops, and Particles, Academic Press, New York.

Corn, M. (1966), "Adhesion of' Particles," Published as Chapter XI of Aerosol Science (C.N. Davies, Ed.), Academic Press., London.

Corn, M. and Stein, F. (1965), "Re-entrainment of Particles from a Plane Surface," American Industrial Hygiene Association Journal, July-August, pp. 325–336.

Cowherd, Chatten Jr., (1981), "Control Of Windblown Dust From Storage Piles" Environment International, Vol. 6, pp. 307–311.

Davis, R.H. and Birdsell, K.H. (1988), "Hindered Settling of Semidilute Monodisperse and Polydisperse Suspensions," AIChE Journal, v. 314, N.1, pp. 123–129.

Dawes, J. G. (1952) "Dispersion of Dust Deposits by Blasts of Air," SMRE Research Reports, No. 36 and 149.

Dhodapkar, Shrikant V., George E. Klinzing (1993), "Pressure Fluctuations In Pneumatic Conveying Systems," Powder Technology, Volume 74, Issue 2, February, pp. 179–195.

Eadie, F.S. and Payne, R.E. (1956), "The Micromerograph - A New Instrument for Particle Size Distribution Analysis," British Chemical Engineering, v. 1, pp. 306–311.

Eckhoff, R. K. (1999). Dust explosions in the process industries. Butterworh-Heineman.

Edwards, John C. and Kathleen M. Ford (1988), "Model of Coal Dust Explosion Suppression by Rock Dust Entrainment," US Bureau of Mines, RI 9206.

Eric S. Weiss, Kenneth L. Cashdollar, Michael J. Sapko, and Eugene M. Bazala (1995), "Secondary Explosion Hazards During Blasting in Oil Shale and Sulfide Ore Mines," United States Bureau Of Mines, RI 9632.

Faeth, G.M. (1987), "Mixing, Transport and Combustion in Sprays," Prog. Energy Combust. Sci., v. 13, pp. 293–345.

Fletcher, B. (1976), "The Interaction of a Shock with a Dust Deposit," J. Phys. D: Appl. Phys., v. 9, pp. 197–202.

Friess, H. and Yadigaroglu, G. (2002), "Modelling of the Resuspension Of Particle Clusters From Multilayer Aerosol Deposits With Variable Porosity," Aerosol Science 33, pp. 883–906

Fuchs, N.A. (1964), The Mechanics of Aerosols, The MacMillan Company, New York.

Geldart, D. (1973), "Types of Gas Fluidization," Powder Technology, 7, pp. 285–292.

Geldart, D., E.C. Abdullah, A. Verlinden (2009), "Characterisation of Dry Powders," Powder Technology, 190, pp. 70–74.

Gerrard, J. H. (1963), "An Experimental Investigation of the Initial Stages of the Dispersion of Dust by Shock Waves," Brit. J. Appl. Phys., v. 14, pp. 186–192.

Giess, P., A. J. H. Goddard, and G. Shaw (1997), "Factors Affecting Particle Resuspension From Grass Swards," J. Aerosol Sci., v. 28, No. 7, pp. 1331–1349.

Gilette, D.A. (1978), "Tests with a Portable Wind Tunnel for Determining Wind Erosion Threshold Velocities," Atmos. Env. v. 12, pp. 2309–2313.

Gilliesa, J.A., V. Etyemezian, H. Kuhns, D. Nikolic, D.A. Gillette (2005), "Effect of vehicle characteristics on unpaved road dust emissions," Atmospheric Environment, 39, pp. 2341–2347.

Gotoh, K. and H. Masuda (1998), "Enhancement Of Removal Efficiency Of Deposited Single Particles By A High Speed Air Jet," J. Aerosol Sci. Vol. 29. Suppl. I, pp. S1231-S1232.

Gradon, Leon (2009), "Resuspension Of Particles From Surfaces: Technological, Environmental And Pharmaceutical Aspects," Advanced Powder Technology, 20, pp. 17–28.

Grzybowski, Krzysztof and Leon Gradon (2005), "Modeling Of The Re-Entrainment Of Particles From Powder Structures," Advanced Powder Technol., Vol. 16, No. 2, pp. 105–121.

Grzybowski, Krzysztof and Leon Gradon (2007), "Re-entrainment of Particles From Powder Structures: Experimental Investigations," Advanced Powder Technol., Vol. 18, No. 4, pp. 427–439.

Hagen, L.J., S. van Pelt, B. Sharratt (2010), "Estimating the saltation and suspension components from field wind erosion," Aeolian Research, v. 1 pp. 147–153.

Harris, M. L. (2010), Private Communication, Pittsburgh Research Center, NIOSH.

Harris, M L, Cashdollar, K L, Man, C, and Thimons E (2009). Mitigating coal dust explosions in modern underground coal mines, in Proceedings of the 9th International Mine Ventilation Congress, 8 pp (New Delhi, India, November 10–13, 2009).

Hartenbaum, Bruce (1971), "Lofting of Particulates by a High Speed Wind," Defense Nuclear Agency Technical Report DNA-2737 ATR-71-25, September.

Hayden, Kimberly S., Kinam Park, Jennifer S. Curtis (2003), "Effect of Particle Characteristics On Particle Pickup Velocity," Powder Technology, 131, pp. 7–14.

Herbreteau, C., R. Bouard (2000), "Experimental Study Of Parameters Which Influence The Energy Minimum In Horizontal Gas–Solid Conveying," Powder Technology, 112, pp. 213–220.

Hertzberg, M. (1987), "A Critique of of the Dust Explosibility Index: An Alternative for Estimating Explosion Probabilities," U.S. Bureau of Mines, RI 9095.

Hidy, G.M. (2004), Aerosols, Encyclopedia of Physical Science and Technology, pp. 273–299.

Hong, J., Y. Shen, Y. Tomita (1995), "Phase Diagrams In Dense Phase Pneumatic Transport," Powder Technology, 84, pp. 213–219.

Hontanon, E., A. de los Reyes and J. A. Capitao (2000), "The CAESAR Code For Aerosol Resuspension In Turbulent Pipe Flows. Assessment Against The Storm Experiments," J. Aerosol Sci. Vol. 31, No. 9, pp. 1061–1076.

Hwang, C. C. (1986), "Initial Stages Of The Interaction Of A Shock Wave With A Dust Deposit," Int. J. Multiphase Flow, v. 12. No 4, pp. 655–666.

Hwang, C.C., Singer, J.M. and Hartz, T.N. (1974), "Dispersion of Dust in a Channel by a Turbulent Gas Stream," U.S. Bureau of Mines, RI 7854.

Ibrahim, A.H., P.F. Dunn, M.F. Qazi (2008), "Experiments And Validation Of A Model For Microparticle Detachment From A Surface By Turbulent Air Flow," Aerosol Science, 39, pp. 645–656.

Iimura, Kenji, Satoshi Watanabe, Michitaka Suzuki, Mitsuaki Hirota, Ko Higashitani (2009), "Simulation of entrainment of agglomerates from plate surfaces by shear flows," Chemical Engineering Science, 64, pp. 1455–1461.

Ilea, C.G., P. Kosinski, A.C. Hoffmann (2008), "Three-Dimensional Simulation Of A Dust Lifting Process With Varying Parameters," International Journal of Multiphase Flow, 34, pp. 869–878.

Ilea, C.G., P. Kosinski, A.C. Hoffmann (2008a), "Simulation of a dust lifting process with rough walls," Chemical Engineering Science, 63, pp. 3864–3876.

Ilea, Catalin G., Pawel Kosinski, Alex C. Hoffmann (2009), "The Effect Of Polydispersity On Dust Lifting Behind Shock Waves," Powder Technology, 196, pp. 194–201.

Jenike, A.W. (1964), "Storage and Flow of Solids," Utah Engineering Experiment Station, Bulletin No. 123, University of Utah.

Jeremy P. Rishel, Elaine G. Chapman (2008), "An Evaluation Of The Wind Erosion Module in DUSTRAN," Atmospheric Environment, 42, pp. 1907–1921.

Jiang, Yanbin, Shuji Matsusaka, Hiroaki Masuda, Yu Qian (2008), "Characterizing The Effect Of Substrate Surface Roughness On Particle–Wall Interaction With The Airflow Method," Powder Technology, 186, pp. 199–205.

Kalman, Haim, Andrei Satran, Dikla Meir, Evgeny Rabinovich (2005), "Pickup (Critical) Velocity Of Particles," Powder Technology, 160, pp. 103–113.

Kauffman, C. W., M. Sichel and P. Wolanski (1992), "Research On Dust Explosions at the University of Michigan," Powder Technology, 71, pp. 119–134.

Kauffman, C.W. (1987), "Recent Dust Explosion Experiences in the U.S. Grain Industry," Industrial Dust Explosions, ASTM STP 958, K.L. Cashdollar, and H. Hertzberg, Eds., American Society for Testing and Materials, Philadelphia, pp. 243—264.

Klemens, R., P. Zydak, M. Kaluzny, D. Litwin, P. Wolanski (2006), "Dynamics of Dust Dispersion From The Layer Behind The Propagating Shock Wave," Journal of Loss Prevention in the Process Industries, 19, pp. 200–209.

Klinzing, George E. (1979), "Influences of Metal Surfaces On The Adhesion Of Coal," Fuel, Volume 58, Issue 11, November, pp. 831–833.

Klinzing, George E. (2004), "Pneumatic Transport," Encyclopedia of Physical Science and Technology, Pages 509–520.

Konrad, K. (1986), "Dense-Phase Pneumatic Conveying : A Review," Powder Technology, 49, pp. 1–35.

Kosinski, Pawel, Alex Christian Hoffmann, Rudolf Klemens (2005), "Dust Lifting Behind Shock Waves: Comparison Of Two Modelling Techniques" Chemical Engineering Science, 60, pp. 5219–5230.

Kosugi, Kenji, Takeshi Sato, Atsushi Sato (2004), "Dependence of drifting snow saltation lengths on snow surface hardness," Cold Regions Science and Technology, v. 39, pp. 133– 139.

Krantz, W.B., Carley, J.F. and Al-Taweel, A.M. (1973), "Levitation of Solid Spheres in Pulsating Liquids," Ind. Eng. Chem. Fundam., v. 12, No. 4, pp. 391–396.

Kraus, M.N. (1980), "Pneumatic Conveying of Bulk Materials," 2nd Edition, Chemical Engineering, McGraw Hill, New York.

Kuhl, Allen L., R. E. Ferguson, K. -Y. Chien. and J. P. Collins (1992), "Turbulent Dusty Boundary Layer in an ANFO Surface-Burst Explosion," Defense Nuclear Agency Technical Report DNA-TR-92-17.

Li, Yintang, Yi Guo (2008), "Numerical Simulation Of Aeolian Dusty Sand Transport In A Marginal Desert Region At The Early Entrainment Stage," Geomorphology, 100, pp. 335–344.

Li, Yu-Chen, C. William Kauffman, and Martin Sichel (1995), "An Experimental Study Of Deflagration To Detonation Transition Supported By Dust Layers," Combustion And Flame, 100, pp. 505–515.

Lock, G. D. (1994), "Experimental Measurements In A Dusty-Gas Shock Tube," Int. J. Multiphase Flow, Vol. 20, No. 1, pp. 81–98.

Lopez, M.V. (1998), "Wind Erosion In Agricultural Soils: An Example Of Limited Supply Of Particles Available For Erosion," Catena 33, pp. 17–28.

Louey, Margaret D., Michiel Van Oort, Anthony J. Hickey (2006), "Standardized Entrainment Tubes For The Evaluation Of Pharmaceutical Dry Powder Dispersion," Aerosol Science, 37, pp. 1520–1531.

Lu, Shou-Xiang, Zi-Ru Guo, Yuan-Long Li, Wei-Cheng Fan, Li Zhang, Li-Zhong Yang and Qing-An Wang (2002), "Experimental And Theoretical Analysis Of Acceleration Of A Gas Flame Propagating Over A Dust Deposit," Proceedings of the Combustion Institute, Volume 29, pp. 2839–2846.

Lundgren, D.A. and Rangaraj, C.N. (1986), "A Method for the Estimation of Fugitive Dust Emission Potentials," Powder Technology, v. 47, pp. 61–69.

Masuda, Hiroaki, Kuniaki Gotoh, Hiroshi Fukuada and Yoshiji Banba (1994), "The removal of particles from flat surfaces using a high-speed air jet," Advanced Powder Technol., Vol. 5, No. 2, pp. 205–217.

Merrison, J.P., H.P. Gunnlaugsson, P. Nørnberg, A.E. Jensen, K.R. Rasmussen (2007), "Determination Of The Wind Induced Detachment Threshold For Granular Material On Mars Using Wind Tunnel Simulations," Icarus, 191, pp. 568–580.

Merrison, J.P., H. Bechtold, H. Gunnlaugsson, A. Jensen, K. Kinch, P. Nornberg, K. Rasmussen (2008), "An Environmental Simulation Wind Tunnel For Studying Aeolian Transport On Mars," Planetary and Space Science, 56, pp. 426–437.

Michelis, J., Margenburg, B., Muller, G. and Kleine, W. (1987), "Investigations into the Buildup and Development Conditions of Coal Dust Explosions in a 700—rn Underground Gallery," Industrial Dust Explosions, ASTM STP 958, K.L. Cashdollar and H. Hertzberg, Eds., American Society for Testing and Materials, Philadelphia, pp. 124–137.

Mirels, H. (1984), "Blowing Model for Turbulent Boundary-Layer Dust Ingestion," AIAA Journal, v. 22, No. 11, PP. 1582–1589.

Mirels, H. (1986), "Blowing Model for Turbulent Boundary-Layer Dust Ingestion," The Aerospace Corporation Report SD-TR-85–97, 10 February.

Mollinger, A.M., F.T.M. Nieuwstadt, J.C.M. Marijnissen, I. B. Scarlett (1992), "Entrainment Of Particles In A Turbulent Boundary Layer, Evaluation Of The Forces On A Single Particle," J. Aerosol Sci., Vol. 23, Suppl. 1, pp. S47-S50.

NMAB (1982), "Pneumatic Dust Control in Grain Elevators: Guidelines for Design, Operation and Maintenance," National Materials Advisory Board, U.S. NRC/NAS, Publication NMAB 367–3, Available from NTIS.

O'Neill, M.E. (1968), "A Sphere in Contact with a Plane Wall in a Slow Linear Shear Flow," Chemical Engineering Science, v. 23, pp. 1293–1298.

Okamoto, S. (1979), "Turbulent Shear Flow Behind Sphere Placed on Plane Boundary," Proc. 2nd Symp. on Turbulent Shear Flows, London, pp. 16.1–16.6.

Okina, G.S., D.A. Gillette, J.E. Herrick (2006), "Multi-scale controls on and consequences of aeolian processes in landscape change in arid and semi-arid environments," Journal of Arid Environments, 65, pp. 253–275.

Owen P.R. (1964), "Saltation of Uniform Grains in Air," J. Fluid Mech, v. 20, Part 2, pp. 225–242.

Owen, P.R. (1969), "Pneumatic Transport," J. Fluid Mech., v. 39, Part 2, pp. 407–432.

Phillips, M. (1980), "A Force Balance Model for Particle Entrainment into a Fluid Stream," J. Phys. D: Appl. Phys., v. 13, pp. 221–233.

Piccinini, Norberto (2008), "Dust Explosion In A Wool Factory: Origin, Dynamics And Consequences," Fire Safety Journal, 43, 189–204.

Quan, Victor (1972), "Couette Flow With Particle Injection," Int. J. Heat Mass Transfer. Vol. 15, pp. 2173–2186.

R. A. Gaj and R. D. Small (1991), "Target Area Operating Conditions - Dust Lofting from Natural Surfaces," Defense Nuclear Agency Technical Report DNA-TR-90–71, June.

Rabinovich, Evgeny, Haim Kalman (2007), "Pickup, Critical And Wind Threshold Velocities Of Particles," Powder Technology, 176, pp. 9–17.

Rabinovich, Evgeny, Haim Kalman (2008), "Boundary Saltation And Minimum Pressure Velocities In Particle–Gas Systems," Powder Technology, 185, pp. 67–79.

Rabinovich, Evgeny, Haim Kalman (2009), "Incipient Motion Of Individual Particles In Horizontal Particle–Fluid Systems: A. Experimental Analysis," Powder Technology, Volume 192, Issue 3, 25 June, pp. 318–325.

Rabinovich, Evgeny, Haim Kalman (2009), "Incipient Motion Of Individual Particles In Horizontal Particle–Fluid Systems: B. Theoretical Analysis," Powder Technology, 192, pp. 326–338.

Rabinovich, Evgeny, Haim Kalman (2009), "Pickup Velocity From Particle Deposits," Powder Technology, 194, pp. 51–57.

Rabinovich, Evgeny, Haim Kalman (2010), "Phenomenological Study Of Saltating Motion Of Individual Particles In Horizontal Particle–Gas Systems," Chemical Engineering Science, Volume 65, Issue 2, 16 January, pp. 739–752.

Rae, D. (1973), "Initiation Of Weak Coal-Dust Explosions In Long Galleries And The Importance Of The Time Dependence Of The Explosion Pressure," 14th Symposium (Int.) on Combustion

Ranade, M.B. (1987), "Adhesion and Removal of Fine Particles on Surfaces," Aerosol Science and Technology, v. 7, pp. 161–176.

Rasmussen, Keld R., Jasper F. Kok, Jonathan P. Merrison (2009), "Enhancement In Wind-Driven Sand Transport By Electric Fields," Planetary and Space Science, 57, pp. 804–808.

Rastogi, S., S. V. Dhodapkar, F. Cabrejos, J. Baker, M. Weintraub, G. E. Klinzing and W.C. Yang (1993), "Survey of characterization Techniques Of Dry Ultrafine Coals And Their Relationships To Transport, Handling And Storage," Powder Technology, 74, pp. 47–59.

Richmond, J.K. and Liebman, I. (1974), "A Physical Description of Coal Mine Explosions," Fifteenth Symposium (International) on Combustion, The Combustion Institute, Pittsburgh, pp. 115–126.

Rodgers, S.A. and Ural, E.A. (2010), "Practical Issues with Marginally Explosible Dusts - Evaluating the Real Hazard," Accepted for publication in the Process Safety Progress.

Roney, Jason A., Bruce R. White (2006), "Estimating Fugitive Dust Emission Rates Using An Environmental Boundary Layer Wind Tunnel," Atmospheric Environment, 40, pp. 7668–7685.

Rubinow, S.I. and Keller, J.B. (1961), "The Transverse Force on a Spinning Sphere Moving in a Viscous Fluid," J. Fluid Mech., v. 11, pp. 447–459.

Rumpf, H. (1977), "Particle Adhesion," Agglomeration 77: Proc. 2nd Int. Symp. on Agglomeration, (K.V.S. Sastry, Ed.), American Institute of Mining, Metallurgical, and Petroleum Engineers, New York, v. 1, pp. 97–129.

Saffman, P.G. (1965), "The Lift on a Small Sphere in a Slow Shear Flow," originally published in J. Fluid Mech., v. 22, Part 2, pp. 385–400, and amended in v. 31, p. 624.

Sanchez, Luis, Nestor Vasquez, George E. Klinzing and Shrikant Dhodapkar (2003), "Characterization of Bulk Solids To Assess Dense Phase Pneumatic Conveying," Powder Technology, v. 138, Issues 2–3, 10 December, pp. 93–117.

Sapko, Michael J., Eric S. Weiss and Richard W. Watson (1987), "Explosibility of Float Coal Dust Distributed over a Coal-Rock Dust Substratum," Proceedings of the 22nd International Conference of Safety in Mines Research Institutes, 20 May.

Savage, Stuart B., Robert Pfeffer, Zhong M. Zhao (1996), "Solids Transport, Separation And Classification," Powder Technology, 88, pp. 323–333.

Scherpa, Thomas (2002), "Secondary Dust Cloud Formation From An Initiating Blast Wave," Fire Protection Engineering Masters Thesis, Worcester Polytechnic Institute, January.

Sehmel, G. A., "Particle Resuspension: A Review," Environment International, Vol. 4, pp. 107–127.

Shaw, William J., K. Jerry Allwine, Bradley G. Fritz, Frederick C. Rutz, Singer, J.M., Cook, E.B. and Grumer, J. (1972), "Dispersal of Coal- and Rock-Dust Deposits," U.S. Bureau of Mines, RI 7642.

Singer, J.M., Greninger, N.B. and Grumer, J. (1969), "Some Aspects of the Aerodynamics of Formation of Float Coal Dust Clouds," U.S. Bureau of Mines, RI 7252.

Singer, J.M., Harris, M.E. and Grumer, J. (1976), "Dust Dispersal by Explosion—Induced Airflow: Entrainment by Airblast," U.S. Bureau of Mines, RI 8130.

Singer, J.M., M.E. Harris, and J. Grumer (1976), "Dust Dispersal by Explosion-Induced Airflow: Entrainment by Airblast," US Bureau of Mines, RI 8130, 1976

Skjold, T., Eckhoff, R.K., Arntzen, B.J., Lebecki, K., Dyduch, Z., Klemens, R., & Zydak, P. (2009), "Simplified Modelling Of Explosion Propagation By Dust Lifting In Coal Mines," 5th ISFEH.

Skjold, Trygve (2007), "Simulating the Effect of Release of Pressure and Dust Lifting on Coal Dust Explosions," 21st ICDERS July 23–27, Poitiers, France.

Smith II, W.J., Whicker, F.W., and Meyer, H.R. (1982), "Review and Categorization of Saltation, Suspension, and Resuspension Models," Nuclear Safety, v. 23, No. 6, pp. 685–699.

Srinath, S.R.., Kaufman, C.W., Nicholls, J.A. and Sichel, M. (1987), "Secondary Dust Explosions," Industrial Dust Explosions, ASTM STP 958, K.L. Cashdollar, and M. Hertzberg, Eds., American Society for Testing and Materials, Philadelphia, pp. 90–106.

Stevenson, P., R. B. Thorpe, J. F. Davidson (2002), "Incipient motion of a small particle in the viscous boundary layer at a pipe wall," Chemical Engineering Science, 57, pp. 4505–4520.

Tadmor, J. And I. Zur (1981), "Resuspension Of Particles From A Horizontal Surface," Atmospheric Environment, Vol. 15 pp. 141–149.

Tamanini, F. (1983), "Dust Explosion Propagation in Simulated Grain Conveyor Galleries," FMRC Technical Report, J.I. OF1R2.RK, July 1983.

Tamanini, F. and E. A. Ural (1992), "FMRC Studies Of Parameters Affecting The Propagation Of Dust Explosions," Powder Technology, 71, pp. 135–151.

Traoré, M., O. Dufaud, L. Perrin, S. Chazelet, D. Thomas (2009), "Dust Explosions: How Should The Influence Of Humidity Be Taken Into Account?," Process Safety And Environment Protection, 87, pp. 14–20.

Trygve Skjold (2007), "Review of the DESC Project," Journal of Loss Prevention in the Process Industries, 20, pp. 291–302.

Ural, E.A. (1989), "Dispersibility of Dusts Pertaining to their Explosion Hazard," FMRC Technical Report J.I. 0Q2E3.RK.

Ural, E.A. (1991), "Experimental Measurement of the Aerodynamic Entrainability of Dust Deposits," Dynamics of Deflagrations and Reactive Systems: Heterogeneous Combustion, Progress in Astronautics and Aeronautics, Vol. 132, pp. 73–92, 1991.

Ural, E.A. (1992), "Dust Entrainability and its Effect on Explosion Propagation in Elongated Structures," Plant/Operations Progress, v. 11, no. 3, pp. 176–181.

Ural, E.A. (1992), "Dust Entrainability and its Effect on Explosion Propagation in Elongated Structures," AIChE 1992 Spring National Meeting, Characterization of the Hazard Potential of Chemicals.

VDI 2263 Part 9: Dust Fires And Dust Explosions: Hazards–Assessment–Protective Measures, "Determination Of Dustiness Of Bulk Materials," May 2008

Venkatasubramanian, Srikanth, George E. Klinzing, Brian Ence (2000), "Flow Rate Measurements Of A Fibrous Material Using A Pressure Drop Technique," Flow Measurement and Instrumentation, Volume 11, Issue 3, September, pp. 177–183.

Villareal, Jesus A., George E. Klinzing (1994), "Pickup Velocities Under Higher Pressure Conditions," Powder Technology, Volume 80, Issue 2, August, pp. 179–182.

Vogl, A. (2010), Private Communication, FSA, Germany.

Witze, P.O. and Dwyer, H.A. (1976), "The Turbulent Radial Jet," J. Fluid Mech., v. 75, part 3, pp. 401–417.

Wypych, Peter W. (1998), "Design Considerations of Long-Distance Pneumatic Transport and Pipe Branching Fluidization," Solids Handling, and Processing, pp. 712–772.

Wypych, Peter W. (1999), "Pneumatic Conveying Of Powders Over Long Distances And At Large Capacities" Powder Technology, Volume 104, Issue 3, October, pp. 278–286.

Wypych, Peter, Dave Cook, Paul Cooper (2005), "Controlling dust emissions and explosion hazards in powder handling plants Chemical Engineering and Processing," 44, 323–326.

Yaping Shao, Yan Yang (2005), "A scheme for drag partition over rough surfaces," Atmospheric Environment, 39, pp. 7351–7361.

Yoshida, T., Kousaka, Y. and Okuyarna, K. (1979), Aerosol Science for Engineers, Power Co., Ltd., Tokyo.

Zhang, Wei, Jong-Hoon Kang, Sang-Joon Lee (2007), "Tracking Of Saltating Sand Trajectories Over A Flat Surface Embedded In An Atmospheric Boundary Layer," Geomorphology, 86, pp. 320–331.

Zimon, A. D. (1982), Adhesion of Dust and Powder, Second Edition, Consultants Bureau, New York.

Ziskind, G., M. Fichman and C. Gutfinger (1995), "Resuspension Of Particulates From Surfaces To Turbulent Flows-Review And Analysis," J. Aerosol Sci., Vol. 26, No. 4, pp. 613–644.

Zydak, P., R. Klemens (2009), "Experimental Investigation into Coal Dust Lifting Process Behind Shock Wave for Different Dust Layer Thicknesses," 5th ISFEH

Zydak, Przemyslaw, Rudolf Klemens (2007), "Modelling Of Dust Lifting Process Behind Propagating Shock Wave," Journal of Loss Prevention in the Process Industries, 20, pp. 417–426.